22nd Journees de la Neutronique 2014 (JDN22)

Crystallography and Neutrons

EPJ Web of Conferences Volume 155 (2017)

Oleron, France
21 - 24 September 2014

Editors:

J.M. Kiat
P.E. Janolin
B. Gillon
F. Porcher
M.T. Fernandez-Diaz

ISBN: 978-1-5108-5052-1

Printed from e-media with permission by:

Curran Associates, Inc.
57 Morehouse Lane
Red Hook, NY 12571

Some format issues inherent in the e-media version may also appear in this print version.

Printed by Curran Associates, Inc. (2017)

For additional information, please contact EDP Sciences – Web of Conferences at the address below.

EDP Sciences – Web of Conferences
17, Avenue du Hoggar
Parc d'Activité de Courtabœuf
BP 112
F-91944 Les Ulis Cedex A
France

Phone: +33 (0) 1 69 18 75 75
Fax: +33 (0) 1 69 28 84 91

contact-edps@webofconferences.org

Additional copies of this publication are available from:

Curran Associates, Inc.
57 Morehouse Lane
Red Hook, NY 12571 USA
Phone: 845-758-0400
Fax: 845-758-2633
Email: curran@proceedings.com
Web: www.proceedings.com

TABLE OF CONTENTS

edp sciences

EPJ Web of Conferences

Statement of Peer review

In submitting conference proceedings to EPJ Web of Conferences, I certify to the Publisher that I adhere to the **Policy on Publishing Integrity** of the journal in order to safeguard good scientific practice in publishing.

1. All articles have been subjected to peer review administered by the proceedings editors.
2. Reviews have been conducted by expert referees, who have been requested to provide unbiased and constructive comments aimed, whenever possible, at improving the work.
3. Proceedings editors have taken all reasonable steps to ensure the quality of the materials they publish and their decision to accept or reject a paper for publication has been based only on the merits of the work and the relevance to the journal.

Title, date and place of the conference

JDN 22 - Crystallography and Neutrons

21 - 26 September 2014, Oléron Island, FRANCE

Proceedings editor(s):

J. M. Kiat, P. E. Janolin (SPMS, Ecole Centrale Paris)

B. Gillon, F. Porcher (LLB, CEA/Saclay)

M. T. Fernandez-Diaz (ILL, Grenoble)

Date and editor's signature

26/09/2017

Foreword

The international thematic school on "Crystallography and Neutrons" was organized in order to celebrate the International Year of Crystallography in 2014, under the patronage of the "Société Française de la Neutronique" (SFN) with the support of the CNRS and the European neutron facilities network NMI3[1]. The school took place in September 2014 in Oléron, France, as first part of the "22ème Journées de la Neutronique" (JDN22). The school gathered participants from European countries and therefore the lectures were given in English. It intended to promote cooperation and networking among users of large scale facilities with researchers involved in the field of Crystallography.

Crystallography is fundamental in Physics, Chemistry and Materials Science because it permits to link the structural and dynamic properties at the atomic scale (also accessible by theoretical calculations) to the local and macroscopic properties (observed by other experimental probes). Contrary to a usual belief, crystallography is not limited to structure determinations by diffraction and indeed almost all previous thematic SFN schools were devoted to a domain belonging to crystallography. The school "Crystallography and Neutrons" was mainly dedicated to the study of crystalline materials in the solid state using neutron diffraction but also provided introductive lectures on other neutron scattering techniques of interest for condensed matter research.

The school highlighted the specific strengths of neutron scattering (magnetism, light elements, contrast methods) for the study of crystalline materials of various kinds, with regards to other investigative techniques. The most relevant neutron techniques for studying crystalline matter were presented during the school (diffraction, 3-axes spectroscopy, reflectometry), as well as their applications to various problems (nuclear structure / dynamics / magnetism) and various studied materials (powder / single crystal / thin film). Tutorials on data analysis with the Jana and Fullprof softwares were given by their own authors, V. Petříček (Prag) and J. Rodriguez-Carvajal (ILL), for some representative systems, in order to complement the theoretical training.

Reminders on the basics of crystallography, neutron scattering and production of neutron beams were presented at the beginning of the school but unfortunately no written contributions were provided by the lecturers. However, the description of fundamentals of crystallography and neutron-matter interaction as well as the description of neutron sources may be found in previous editions of the SFN collection[2,3]. The principles of the determination of nuclear and magnetic structures from experimental neutron diffraction data were also presented in previous volumes[2,4].

This book gathers lectures going further than routine nuclear or magnetic structure determinations by neutron diffraction. The first chapter provides an overview of the available

[1] Integrated Infrastructure Initiative for Neutron Scattering and Muon Spectroscopy.

[2] JDN 15 - Études structurales par diffraction des neutrons, série des «Écoles Thématiques de La Société Française de la Neutronique», O. Isnard, Ed. (EDP Sciences, Les Ulis, ISBN: 978-2-7598-0407-8,2000).

[3] JDN 10 - Neutrons et Matériaux, W. Paulus et J. Meinnel, J. Phys. IV (Proceedings) **103** (2003), ISBN: 2-86883-656-9.

[4] JDN 20 - Neutrons et Magnétisme, série des «Écoles Thématiques de La Société Française de la Neutronique», V. Simonet, B. Canals, J. Robert, S. Petit et H. Mutka (Eds.) (EDP Sciences, Les Ulis, ISBN: 978-2-7598-1248-6).

neutron diffractometers for powder and single crystal diffraction and more sophisticated instruments using polarized neutron beams, with or without polarization analysis (N. Qureshi, ILL). Twinning and modulation in crystal structure determination by neutron diffraction are examined in the second chapter (V. Petříček and M. Dušek, Prag). The next chapter deals with the structure and dynamics of aperiodic crystals (C. Mariette et al., Rennes). Chapter 4 shows the importance of neutron diffraction for studying texture in material science (V. Klozek, LLB). The last two chapters introduce two worthy techniques for condensed matter physics: diffuse neutron scattering (I. Mirebeau, LLB) which provides insights on nuclear and magnetic short range ordering and inelastic neutron scattering (S. Petit, LLB) which permits to explore lattice and spin dynamics.

We would like to thank all the speakers and participants for the success of this school, as well as our sponsors.

The organizers,
J.M. Kiat, P.E. Janolin (Ecole Centrale-Supelec),
B. Gillon, F. Porcher (Laboratoire Léon Brillouin, CEA-CNRS),
M.T. Fernandez-Diaz (Institut Laue Langevin)

Instrumental aspects

Navid Qureshi

Institut Laue Langevin, 38000 Grenoble, France

Abstract. Every neutron scattering experiment requires the choice of a suited neutron diffractometer (or spectrometer in the case of inelastic scattering) with its optimal configuration in order to accomplish the experimental tasks in the most successful way. Most generally, the compromise between the incident neutron flux and the instrumental resolution has to be considered, which is depending on a number of optical devices which are positioned in the neutron beam path. In this chapter the basic instrumental principles of neutron diffraction will be explained. Examples of different types of experiments and their respective expectable results will be shown. Furthermore, the production and use of polarized neutrons will be stressed.

1. Introduction

A successful neutron scattering experiment requires a thorough plan of which instrument to use and how to set up its configuration in order to obtain the best possible results. Optimally, the decision of the set-up is taken beforehand, but in many cases the instrument parameters are modified on-the-fly after analyzing the first results. Due to the different experimental tasks – reaching from magnetic or nuclear structure investigation to the mapping of phase diagrams – and the different sample dimensions (both the crystal size and the lattice parameters), which can be investigated at one and the same instrument, it becomes obvious that the versatility of some instruments has to be adjusted to satisfy the experimental tasks. Obviously, the sample state (powder, single crystal, liquid, etc.) limits the choice of instruments, however, you can find powder diffractometers which differ from each other concerning the used neutron wavelength and particular devices in the beam path in order to maximize the efficiency for a type of experiment.

In this chapter the different instrumental aspects of powder and single-crystal diffractometers will be discussed including the choice of the sample environment and how this may be tuned in order to influence the obtainable results in terms of intensity, resolution, etc. At the end a short insight into the use of polarized neutrons will be given and how the analysis of the polarization after the scattering process can be exploited for the determination of magnetic properties of single-crystal samples.

2. The basic diffractometer

Regardless of the type of diffractometer (powder, single crystal, high resolution, high flux, etc.) the basic components along the beam path are the same (see Fig. 1) starting with neutrons coming from the source [reactor (continuous), spallation (pulsed)] and passing through optical devices (e.g. collimators, slits, diaphgrams) until a single wavelength is selected at

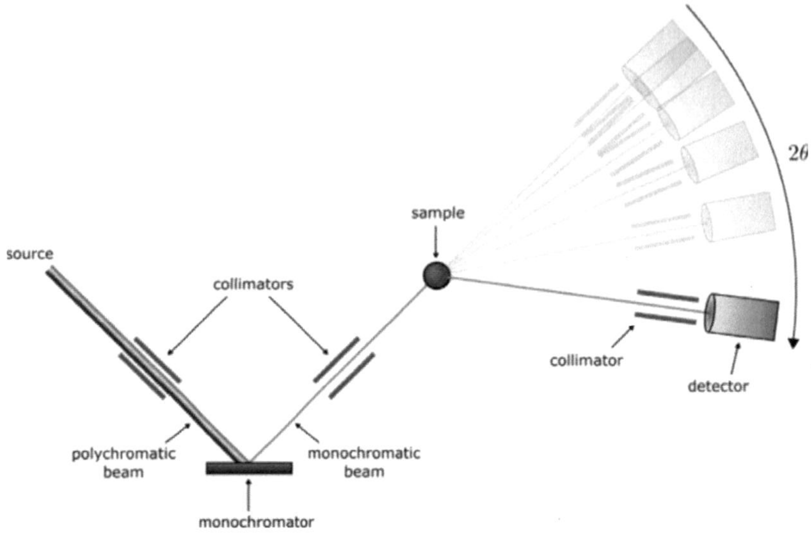

Figure 1. Sketch of a basic diffractometer with the usual components along the beam path.

the monochromator according to the Bragg's law (Eq. (1)).

$$n\lambda = 2d\,sin\theta \tag{1}$$

The monochromatic beam interacts with the sample and is scattered under an angle of 2θ, where it is finally detected. Collimators can be used between (a) the source and the monochromator, (b) the monochromator and the sample and (c) the sample and detector in order to decrease the primary, secondary and tertiary divergence, respectively. A (Soller) collimator for neutron beams consists of parallel sheets made out of or covered with a neutron-absorbing material which only lets pass those neutrons which have a small divergence from the beam axis. In order to adjust the shape and size of the beam at the sample position a set of slits and diaphragms can be used in addition to the collimators. Collimators and slits between the sample and detector are used in order to reduce the background which originates from neutrons being scattered under a different angle at a different position (e.g. at the walls of the sample environment such as a cryostat). Obviously, all optical devices which reduce the beam divergence and affect its size will increase the resolution of the diffractometer. But as this is at the cost of those neutrons which have been absorbed the intensity of neutrons delivered to the sample or to the detector decreases, which constitutes the compromise between flux and resolution.

High quality single crystals or an assembly of those are used as monochromators where the selected wavelength will be diffracted under the monochromator take-off angle $2\theta_M$ which plays an important role in the resolution properties of the diffractometer which will be discussed later. Naturally, the highest possible intensity of neutrons scattered at the monochromator should be used as a primary beam. Therefore, monochromators should be made of a low-absorbing and high-reflecting material, they should provide a finite mosaicity in order to diffract a not too narrow wavelength band, and finally a Bragg reflection with a strong nuclear structure factor should be selected. Further measures can be taken to increase the intensity, however, at the same time they reduce the resolution, which will be explained in detail later. Typical materials for monochromators are Cu, Ge, Si or graphite. According to the Bragg's law [Eq. (1)] for $n > 1$ higher harmonics of the selected wavelength, i.e. $\lambda/2$, $\lambda/3$, ..., are also diffracted at the monochromator for which filters are typically used between the

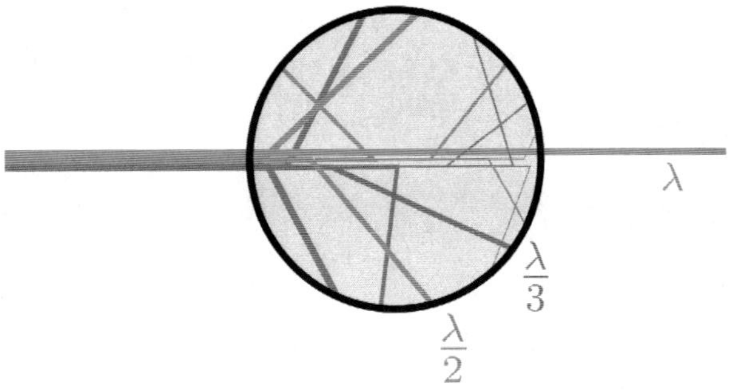

Figure 2. Schematic representation of a higher-harmonics filter.

monochromator and the sample. Such a filter (like pyrolytic graphite or Be) is chosen upon its largest lattice plane distance according to the relation $\lambda < 2d_F$, which states the cut-off wavelength for Bragg diffraction. In such a way all wavelengths inferior to $2d_F$ are diffracted out of the beam path and absorbed at the container walls of the filter (see Fig. 2). For Si or Ge monochromators, typically the (111) reflection is used for monochromatizing the beam, because the (222) reflection is forbidden by the extinction rules of this particular space group.

Now that a monochromatic neutron beam is available, which possibly has been collimated and shaped, the experimental conditions have to be set. Measuring physical properties of a sample at low temperatures (down to the mK region) requires a cryostat (optionally with a $^3He/^4He$ dilution), while for elevated temperatures obviously a furnace is needed. Magnetic fields or pressure can be applied by using a cryomagnet or a pressure cell, respectively. Each of these sample environments has its purpose, but also its limitations in terms of the accessibility of Bragg reflections when single-crystal diffraction is concerned.

Finally, after the interaction with the sample the scattered neutrons travel to the detector (eventually after passing further collimators and slits) where they trigger a reaction with the detector material in order to produce a charged particle which is afterwards accelerated by an electric field and thereby detected. Typically, the ^3He isotope is used as a gas inside the detector cells, because it has a strong neutron absorption cross section and is rather insensitive to gamma radiation. Neutron detectors can consist of a single ^3He cell (single detector), but especially for powder diffraction where the statistical orientation of the crystallites leads to scattering in all possible directions large area detectors are used to cover a large 2θ range. Such an area detector of the D20 diffractometer at the Institut Laue Langevin is shown in Fig. 3.

3. Diffraction techniques

In this section the different diffraction techniques will be explained.

3.1 Powder diffraction

Neutron powder diffraction is a very useful technique for investigating nuclear and magnetic structures because no preliminary structural information of the sample is necessary and relatively fast and easy experiments can be carried out in contrast to single-crystal diffraction as will be shown in the next section. As the vertical divergence is no issue for the resolution of a powder diffractometer sample holders usually are cylindrical in order to have the same amount of powder with less material along the beam path meaning less absorption (Fig. 4). Vanadium is suited as a sample holder material, because it only scatters neutrons incoherently,

Figure 3. Large area detector at the D20 neutron diffractometer at the ILL. ^{3}He gas cells cover an angle of 153.6°. Taken from [1].

Figure 4. Left panel: D20 powder diffractometer at the ILL (taken from [1]). Right panel: Vanadium sample holder.

i.e. it does not produce additional Bragg peaks in the diffraction pattern, but a uniform increase of background.

As mentioned above coherently scattered neutrons will produce Bragg peaks in the diffraction pattern (the scattered intensity as a function of the scattering angle 2θ (see Fig. 1), which bear all the information necessary to determine the crystal or magnetic structure of the investigated material. In this context the observations to be exploited are the positions of the peaks, their intensity as well as their shape and width.

From the position of the Bragg peaks in the diffraction pattern (see Fig. 5), i.e. the scattering angle 2θ, the d value of the lattice plane (hkl) can be calculated by the Bragg's law provided that the wavelength is known. If the diffraction pattern is indexed correctly,

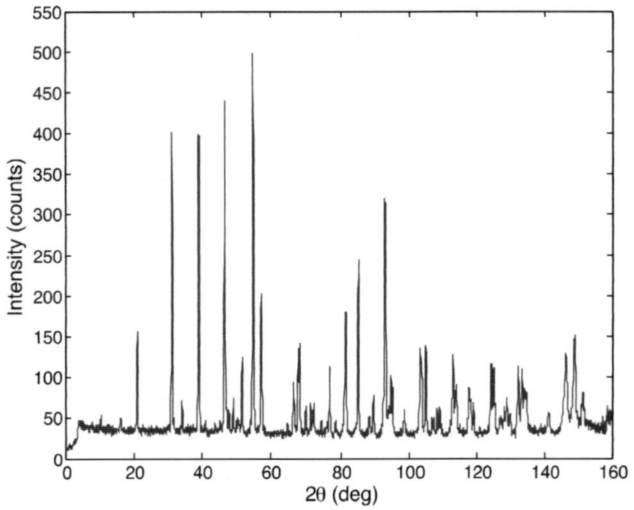

Figure 5. Powder diffraction pattern. Coherently scattered neutrons produce Bragg peaks at certain 2θ values.

i.e. an (*hkl*) plane has been attributed to every Bragg reflection, the lattice parameters of orthorhombic systems (for simplicity) can be obtained from the *d* values according to

$$d = \left(\frac{\boldsymbol{h}^2}{\boldsymbol{a}^2} + \frac{\boldsymbol{k}^2}{\boldsymbol{b}^2} + \frac{\boldsymbol{l}^2}{\boldsymbol{c}^2}\right)^{-\frac{1}{2}} \tag{2}$$

Hence, with the known monochromatic wavelength one is able to obtain the lattice parameters by measuring the scattering angles, which gives the metric of the unit cell. However, this does not yet tell how the atoms are distributed within. This information can be obtained by analyzing the integrated intensity of the Bragg reflections which is proportional to the nuclear [Eq. (3)] or the magnetic structure factor [Eq. (4)] depending on if nuclear or magnetic scattering is concerned.

$$F_N(\boldsymbol{Q}) = \sum_i b_i exp(i\,\boldsymbol{Q}\boldsymbol{r}_i) exp\left(-B_i \frac{sin^2\theta}{\lambda^2}\right) \tag{3}$$

$$\boldsymbol{F_M}(\boldsymbol{Q}) = \sum_i \boldsymbol{\mu}_i f_i(\boldsymbol{Q}) exp(i\,\boldsymbol{Q}\boldsymbol{r}_i) exp\left(-B_i \frac{sin^2\theta}{\lambda^2}\right) \tag{4}$$

Here the sum is taken over all i atoms in the unit cell, b_i is the scattering length, $\boldsymbol{r_i}$ is the vector containing the fractional coordinates and f_i is the magnetic form factor of atom i, with magnetic moment $\boldsymbol{\mu}_i$.

Finally, real structure properties like stress and strain can be obtained from the shape and width of the Bragg reflections. Therefore, in order to derive the effects coming from the sample one must know the instrumental contribution to the peak profile, which depends on many factors, e.g. the size of the neutron source and the monochromator, the system of slits and collimators, the monochromator take-off angle, etc. All these instrument-related factors are expressed with the empirical Cagliotti formula giving the full width at half maximum (FWHM) of a peak as a function of the scattering angle.

$$FWHM^2 = u\tan^2\theta + v\tan\theta + w \tag{5}$$

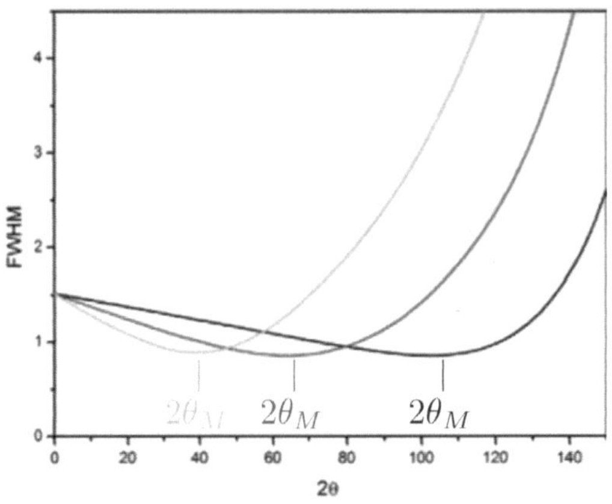

Figure 6. Resolution curve according to the Cagliotti formula for three different arbitrary instrument configurations.

The resulting resolution curve is sketched in Fig. 6 for three different instrument configurations. It has to be pointed out that the resolution function minimum is always located at the scattering angle corresponding to the monochromator take-off angle resulting from a focusing effect combining secondary and tertiary divergence.

Knowing which parts of the diffraction pattern contain the information for which the experiment is carried out at the first place permits the choice and correct configuration of the powder diffractometer. Figure 7 shows powder diffraction data on the oxypnictide iron superconductor $LaO_{1-x}F_xFeAs$ where details both of the crystal and the magnetic structure were examined. In this system a structural phase transition takes place from a tetragonal to an orthorhombic crystal structure upon F-doping or decreasing temperature. The phase transition is manifest in the splitting of the tetragonal (220) reflection into the orthorhombic (400) and (040) reflections due to the splitting of the a and b lattice constants. In order to be able to resolve the double-peak feature in Fig. 7(a) the best possible resolution is needed for which this part of the experiment was carried out at the high-resolution neutron powder diffractometer D2B (ILL). In order to analyze the magnetic structure integrated magnetic Bragg intensities have to be collected which can be related to the magnetic moment alignment according to Eq. (4). Due to the relatively weak intensity of the magnetic reflections (marked by arrows) and due to the fact that the resolution plays a minor role, when integrated intensities have to be measured, such kind of experiment is best done on a high-flux diffractometer. But what exactly are the differences between the two kinds?

First of all, as pointed out above, the resolution curve has its minimum at the monochromator take-off angle $2\theta_M$. Therefore, the investigation of magnetic reflections, which usually occur at small 2θ values due to the magnetic form factor decay, would benefit from a small $2\theta_M$, while a large take-off angle would provide a good resolution over a wide range of the diffraction pattern (see blue curve in Fig. 6). Besides this the take-off angle has another effect. By considering the derivation of the Bragg's law with respect to the scattering angle $\theta = \theta_M$ [Eq. (6)] it can be seen that an increase in the take-off angle goes alongside with a decrease in the selected wavelength band $\Delta\lambda/\lambda$ and vice versa.

$$\frac{\Delta\lambda}{\lambda} = \Delta\theta_M \cot\theta_M \tag{6}$$

Consequently, a narrower wavelength band means less divergence of the beam for which a high take-off angle is generally a good choice for having a high-resolution diffractometer.

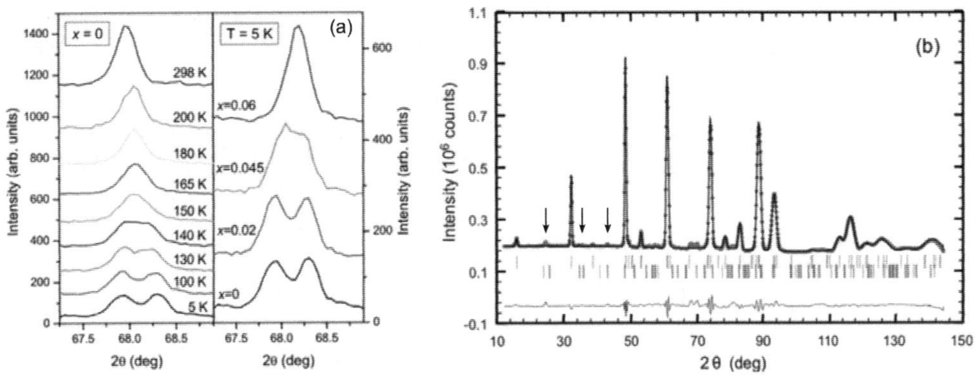

Figure 7. (a) Focus on the $(220)_T$ reflection which splits into the $(400)_O$ and $(040)_O$ reflections at the structural phase transition (measured at the high-resolution diffractometer D2B (ILL). (b) Powder diffraction pattern measured at the high-flux diffractometer D20 (ILL). Weak magnetic reflections indicated by arrows are observed at small diffraction angles. Taken from [1].

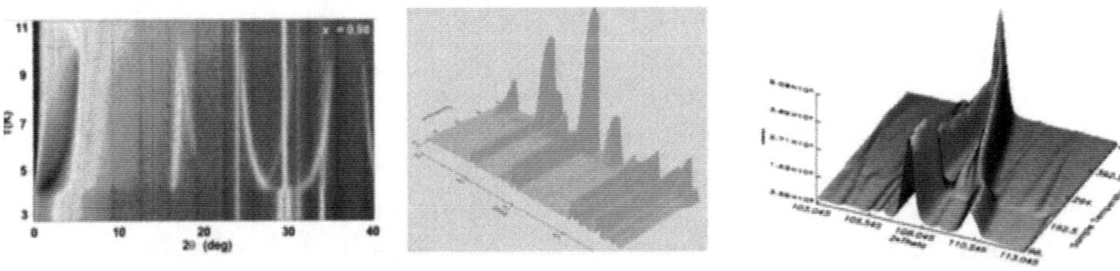

Figure 8. Thermodiffraction as a method to clearly reveal structural or magnetic phase transitions.

To further increase the resolution Soller collimators should be used wherever possible. Especially a collimation before the detector removes the dependence of the resolution on the sample size. On the other hand, if the flux is to be maximized no collimation should be used as this cuts off parts of the neutron beam. Furthermore, a large vertically focusing monochromator with a large mosaic spread increase the flux at the sample where the latter is also related to Eq. (6). As the monochromator reflectivity increases proportionally to λ^3 larger wavelengths provide more flux. Finally, the more sample is in the neutron beam, the higher is the scattered intensity. By optimizing all those parameters a diffraction pattern at the high-flux diffractometer D20 can be recorded in a matter of seconds allowing a collection as a function of e.g. temperature in order to trace phase transitions (see Fig. 8 for examples).

3.2 Single-crystal diffraction

In contrast to powder experiments neutron single-crystal diffraction requires the knowledge of the lattice parameters and the rough orientation of the crystallographic axes within the sample (note that this is not necessary for diffractometers equipped with a very large position-sensitive area detector where hundreds of Bragg reflections can be measured simultaneously permitting to derive the unit cell metrics, e.g. Laue diffractometers or instruments dedicated to fibre diffraction like D19 at the ILL) in order to be able to put the sample in reflection position for a particularly chosen (*hkl*) plane. Due to the necessary alignment of the sample and the individual measurement of (*hkl*) reflections single-crystal diffraction experiments typically last 3–10 days and should only be carried out if neutron powder and X-ray single crystal

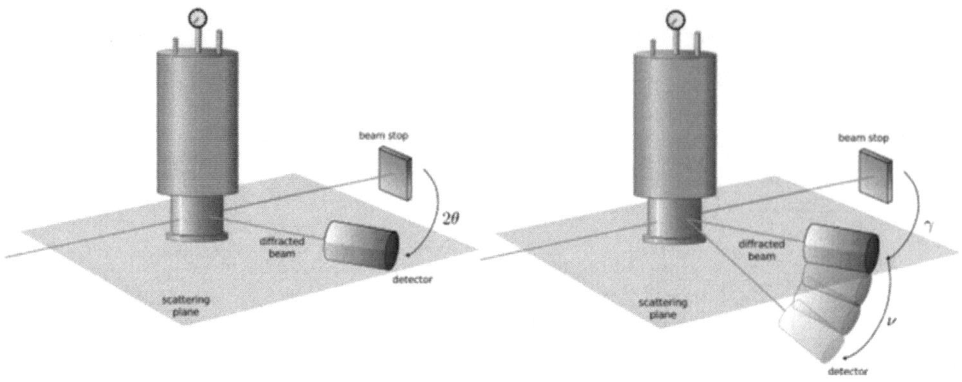

Figure 9. Left panel: Sketch of the normal beam equatorial geometry with the sample within a cryostat. Right panel: Normal beam geometry with lifting arm detector.

experiments fail to provide useful results. Measuring an (*hkl*) reflection is done by positioning the detector at the correct 2θ value (that is why the lattice parameters must be known) and rotating the sample about a vertical axis, which is called the ω angle given that the (hkl) vector to be measured lies within the plane defined by the incident and the scattered beam. This geometry is called the *normal-beam equatorial mode*. When heavy sample environments like cryomagnets are used the tilting angle of the sample is very limited. The *minimal tilt* geometry permits to tilt the sample table with the sample environment by a few degrees. This also means that the sample alignment needs to be quite accurate in order to bring those (*hkl*) vectors which are intended to be measured within the equatorial plane (this is why the orientation of the crystal axes must be known). Another possibility is to move the detector out of the equatorial plane, which is called a *lifting arm* detector and is depicted in the right panel of Fig. 9. For a complete overview of the different single-crystal diffraction geometries, the instrument design and beam optics the interested reader is referred to [3].

The movement of the detector out of the scattering plane is expressed by the angle ν, while the movement within the scattering plane is denoted γ (for $\nu = 0$ γ equals to 2θ). With this option scattering vectors not too far away from the equatorial plane can still be reached and more reflections are available for the subsequent structure refinement.

Sample environments that can be mounted within a Eulerian cradle bear the advantage that the sample can be rotated around two more angles permitting to reach the majority of reciprocal space points within the Ewald sphere. In a so called *4-circle diffractometer* the angle χ rotates the sample around a horizontal axis and the angle ϕ is the rotation around the sample axis (Fig. 10).

In a single-crystal diffraction experiment the sample alignment is almost as important as the data collection itself. The sample should not precess while any of the three angles ω, χ, ϕ is being adjusted while being in the center of the neutron beam. The typical experimental procedure is then to find two linearly independent (*hkl*) reflections which provide the ability to calculate the *orientation matrix* UB [4]. This matrix relates the sample lattice parameters and the diffractometer axes and allows therefore to calculate the diffractometer angles for any (*hkl*) reflection and hence to position the sample accordingly. The UB matrix is usually refined after measuring a set of 20–30 reflections in order to achieve a correct positioning within a large 2θ range and for different regions within the reciprocal lattice [e.g. measuring hundreds of (*hk*0) reflections does not allow the refinement of the UB matrix entries for $l \neq 0$ components]. After this step the experimental conditions like temperature, magnetic field, pressure, etc. are set and – if possible – a few hundreds of reflections are measured at constant

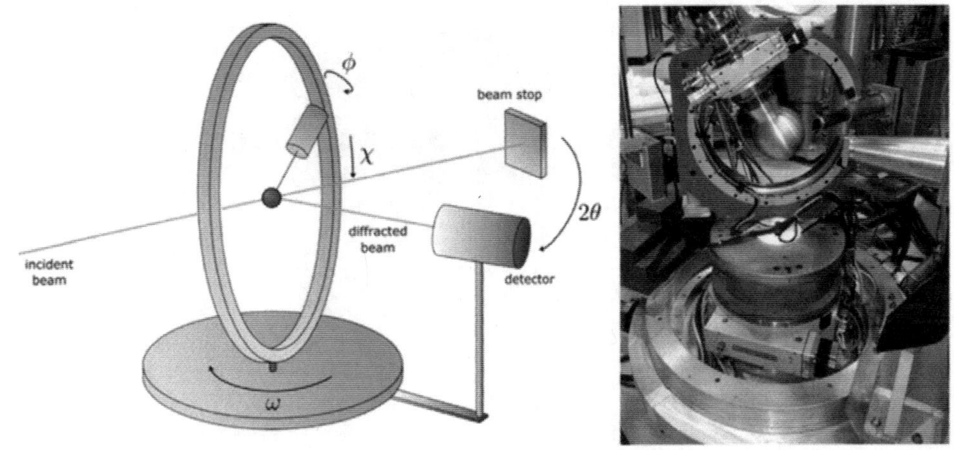

Figure 10. Left panel: Sketch of a Eulerian cradle for an experiment in the 4-circle mode. Right panel: 4-circle cryostat at the D10 diffractometer (ILL).

conditions. Measuring a Bragg reflection consists in a rocking scan around the reflection position, the observed peak is then integrated which yields an observation proportional to the square of the structure factor. In order to improve statistics symmetry-equivalent reflections are merged and averaged. The set of integrated intensities is then used to refine a magnetic or structural model by a least-squares algorithm.

Of course, other types of experiments can be carried out on a single-crystal diffractometer, e.g. phase transitions that manifest themselves by a change in position or intensity of a certain family of (*hkl*) reflections can easily be traced. Sometimes the interesting information is hidden between the Bragg peaks for which a volumetric mapping of the reciprocal space is necessary in order to reveal diffuse scattering and reflection profiles.

3.3 Laue diffraction

The most important feature of Laue diffraction is the use of a white neutron beam, i.e. a large wavelength band is used instead of selecting a wavelength by a monochromator. Furthermore, the single-crystal sample is kept fixed at the center of the Laue diffractometer. All (*hkl*) reflections which lie between the two Ewald spheres with the radii $1/\lambda_{max}$ and $1/\lambda_{min}$ (Ewald construction in reciprocal space) can be measured. In other words due the large wavelength band there is always an adequate wavelength which is able to produce a Bragg peak according to Eq. (1) after being scattered at a randomly oriented (*hkl*) lattice plane. In order to measure a large part of the reciprocal space neutron-sensitive films or cameras can be placed cylindrically around the sample leaving just a small hole for the incoming beam. Laue diffraction allows the quick alignment of single crystals for subsequent structure determination experiments at a single-crystal diffractometer as described in Sect. 2.2. Furthermore, due to the rather fast acquisition of a large portion of reciprocal space magnetic or structural phase transitions can be investigated as a function of temperature (or pressure etc.) by tracing the Laue spots of certain (*hkl*) reflections.

4. Polarized neutrons

The fact that the strength of the interaction between a neutron and the sample is dependent on the orientation of the neutron spin is the principle idea of polarizing neutron beams. The

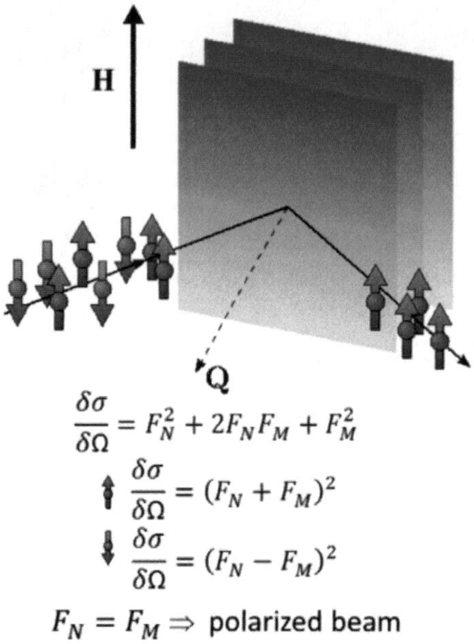

Figure 11. Polarizing crystal producing a polarized neutron beam.

nuclear interaction is only neutron spin dependent, if the scattering nucleus itself has a spin and if the spins of the nuclei are ordered, which only applies to extreme conditions like very low temperatures in the mK region of very high magnetic fields. Therefore, nuclear scattering can be considered as spin independent. In contrast, electron spins are much more easily ordered leading to an important neutron spin dependence of the magnetic scattering. Polarized neutron diffraction is used to determine precise magnetic structure factors by the *flipping ratio* method. Analyzing the polarization state of the neutron after the scattering process offers the possibility to determine the type of interaction and the configuration of magnetic moments in a magnetically ordered compound.

4.1 Production of a polarized neutron beam

Different techniques are used to achieve a neutron beam in which the majority of neutrons have the same spin direction, where the polarization is defined as

$$P = \frac{I_+ - I_-}{I_+ + I_-} \tag{7}$$

Important factors which have to be taken into account are e.g. the desired degree of polarization, the available space at the instrument and the budget.

Polarizing crystals are used as a monochromator, where not only the wavelength is selected, but also the neutron spin state. This is due to the fact that those particular compounds which are used as polarizing monochromators offer a Bragg reflection (*hkl*) where the magnetic structure factor is of the same order as the nuclear structure factor. In order to achieve this the monochromator is placed within a permanent magnetic field **H** which produces a mono-domain state and also helps to tune the magnetic structure factor to the correct value. The spin-dependent nature of the interaction leads to the fact that only one spin state is diffracted as the cross section of the other spin state is zero as can be seen in Fig. 11.

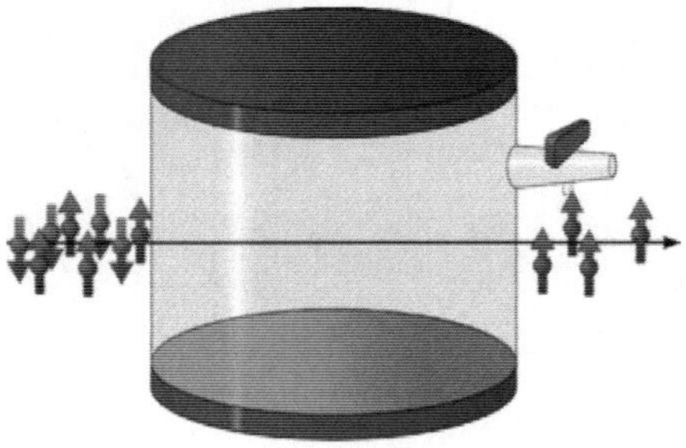

Figure 12. ^{3}He spin filter exploiting the fact that optically pumped ^{3}He ions have a strongly neutron-spin-dependent absorption cross section.

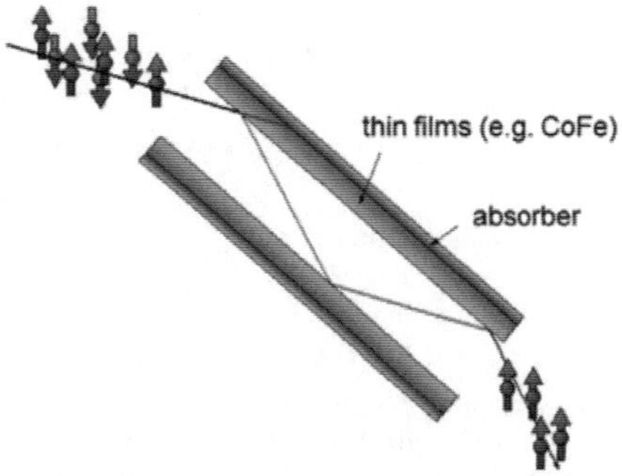

Figure 13. Sketch of a neutron supermirror which polarizes a neutron beam due to the spin-dependent refractive index $\theta_{c\pm}$.

Not only the scattering cross section is spin dependent, but also the absorption cross section. In this context the ^{3}He nucleus turns out to be ideal as it is almost transparent for one neutron spin state and highly absorbing for the other. In such *spin filters* ^{3}He gas under a certain pressure optimizing the transmission-to-absorption ratio is optically pumped by a laser in order to have only one ^{3}He spin state, which will therefore only absorb one of the neutron spin states (Fig. 12).

Another method (among others) is based on the spin-dependent refractive index when neutrons are scattered at thin films of e.g. CoFe under small angles. While e.g. spin-up neutrons undergo total reflection between two limiting walls and therefore get transported, spin-down neutrons penetrate the thin film and are absorbed by the shielding of the guide. A sketch of a so called *neutron supermirror* is shown in Fig. 13.

4.2 Guiding the polarization

Once a neutron beam has been polarized it has to be taken care that the polarization is not affected or lost due to magnetic stray fields. Therefore the neutron polarization has to be

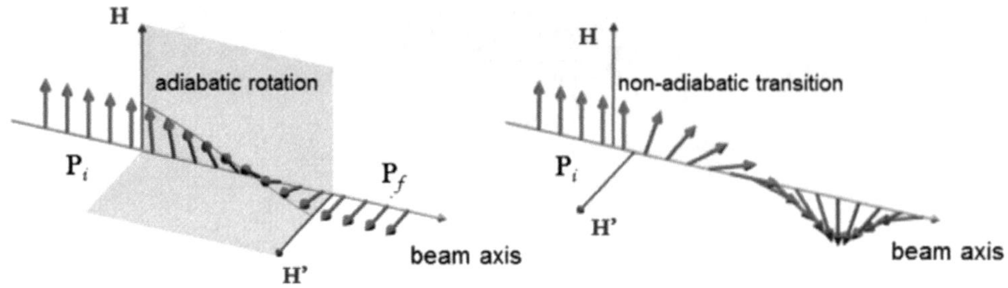

Figure 14. Left: adiabatic rotation of the neutron spin due to the slow enough change of the magnetic field direction. Right: non-adiabatic transition due to the abrupt 90° rotation of the magnetic field. The neutron spin cannot follow the field change and precesses around the new field direction.

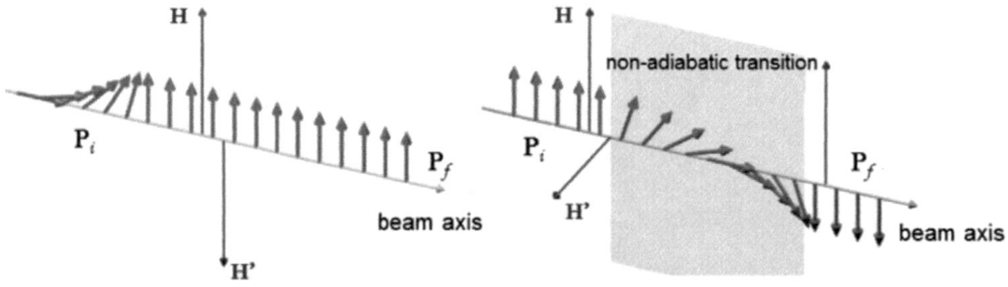

Figure 15. Left: principle of a Cryoflipper. Neutron spins adiabatically rotate towards $\boldsymbol{H}$. Due to the abrupt change from $\boldsymbol{H}$ to $\boldsymbol{H}'$ the neutron spin is effectively flipped with respect to the guide field $\boldsymbol{H}'$. The actual spin flip is effectuated by reversing the direction of $\boldsymbol{H}$. Right: principle of a Mezei flipper. By choosing the correct path length for the non-adiabatic transition the neutron spin can be caught after a 180° rotation by a second vertical magnetic field.

guided along the beam path in so called *guide fields* which are of the order of a few Gauss. Along its travel path the neutron spin precesses around the guide field with the well-defined Larmor frequency ω_L, which is dependent on the strength of the magnetic field **H**:

$$\omega_L = \gamma_N \mu_0 H \tag{8}$$

where γ_{H} is the neutron's gyromagnetic ratio and μ_0 is the vacuum permeability. If the direction of the external magnetic field is varied slowly enough (e.g. as a function of the distance y along the neutron beam path), the neutron spin is able to follow the magnetic field change (see left panel of Fig. 14). An empirical requirement to maintain the polarization is

$$\frac{\omega_L}{\omega_H} > 10 \rightarrow \frac{\partial \theta_H}{\partial y} < 2.65 H\lambda \tag{9}$$

where θ_H is the angle between the magnetic field direction and an arbitrary axis. However, if the change of the magnetic field is too fast for the neutron spin to follow, the latter will precess around the new field direction (see right panel of Fig. 14), which is a non-adiabatic transition.

4.3 Flipping the polarization

Diffraction experiments using polarized neutrons can only supply valuable information if the spin state of the incident neutrons can be controlled in order to reveal the different interaction cross sections with the sample. In some cases the spin state after the scattering has to be

Table 1. Cross sections for polarized neutron scattering with longitudinal polarization analysis ($Q \| x$).

	NSF	SF
$\boldsymbol{P}_0 \| x$	σ_N	$\sigma_{My} + \sigma_{Mz}$
$\boldsymbol{P}_0 \| y$	$\sigma_N + \sigma_{My}$	σ_{Mz}
$\boldsymbol{P}_0 \| z$	$\sigma_N + \sigma_{Mz}$	σ_{My}

manipulated as well. Devices which affect the polarization of a neutron in a way that it is rotated by 180° are called *spin flippers* and use the adiabatic (left panel of Fig. 15) or non-adiabatic transitions (right panel of Fig. 15) explained in the previous section.

4.4 Selection rules for polarized neutrons

Polarized neutrons add additional selection rules to the general law in neutron scattering that only magnetic components perpendicular to the scattering vector contribute. Generally, nuclear scattering is a non-spin-flip (NSF) process regardless of the initial neutron polarization state. Only magnetic components perpendicular to the scattering vector ($\boldsymbol{Q} \parallel \boldsymbol{x}$ by definition) are accessible in a neutron experiment. The components perpendicular to the polarization axis ($\boldsymbol{y}$ being in the scattering plane and z being the perpendicular axis of the instrument) contribute to the spin-flip (SF) channel, while those parallel to the axis of polarization scatter into the NSF channel. The method where the neutron spin is polarized along one axis and then analyzed along the same axis is called *longitudinal polarization analysis (LPA)*. The cross sections for the different polarizations are given in Table 1.

As it can be seen the analysis of the final neutron polarization is a powerful tool as it permits to separate the nuclear and the magnetic cross sections by measuring the appropriate channels. Furthermore, it is possible to gain information about the spatial configuration of the magnetic moments. Such experiments are carried out with a set of Helmholtz coils which generate a magnetic field of a desired direction at the sample position (see Sect. 4.6). Even more information can be acquired when the final polarization is analyzed in all three directions for every individual initial polarization, which is called *spherical polarization analysis (SPA)* and is carried out using the Cryogenic Polarization Device (CryoPAD), see Sect. 4.7.

For more information on this and the next two sections the interested reader is referred to the chapter *Polarized Neutrons and Polarization Analysis* by J. Schweizer in [5].

4.5 Flipping ratio method

The flipping ratio method is used for the experimental derivation of precise magnetic structure factors. Hereby, the interference between nuclear and magnetic scattering and its dependence on the neutron spin are exploited (e.g. $F_M = 0.1 F_N$):

$$I^+ \sim (F_N + F_M)^2 = F_N^2 + 0.2 F_N^2 + 0.01 F_N^2 = 1.21 F_N^2 \tag{10}$$

$$I^- \sim (F_N - F_M)^2 = F_N^2 - 0.2 F_N^2 + 0.01 F_N^2 = 0.81 F_N^2$$

whereas for unpolarized neutrons:

$$I \sim F_N^2 + F_M^2 = F_N^2 + 0.01 F_N^2 = 1.01 F_N^2 \tag{11}$$

Due to the interference term the magnetic structure factor is amplified by the nuclear structure factor. Weak magnetic moments can therefore be measured especially when the nuclear structure factors are strong. The flipping ratio is defined as

$$R = \frac{I^+}{I^-} \tag{12}$$

The collection of a large number of precise magnetic structure factors permits a more sophisticated analysis of the magnetization density by non-spherical models, e.g. the Maximum Entropy Method or the multipole description of the atomic orbitals.

Figure 16 shows the basic setup of a flipping ratio experiment. The neutron beam is polarized and monochromatized at the polarizer and the neutron spin can be flipped by turning the spin flipper on, which permits the measurement of the intensities I^+ and I^-. The sample is placed within an external magnetic field in order to fully saturate the magnetic moments and especially to create a mono-domain state. Therefore, the magnetic moments all point along the vertical axis z. In a flipping ratio experiment the initial polarization of the neutrons is set only along $+z$ or $-z$, i.e. by looking at Table 1 the only cross section to be expected is $\sigma_N + \sigma_{M_z}$ in the NSF channel (as no y component of the magnetic moment is present) and no polarization analysis is necessary.

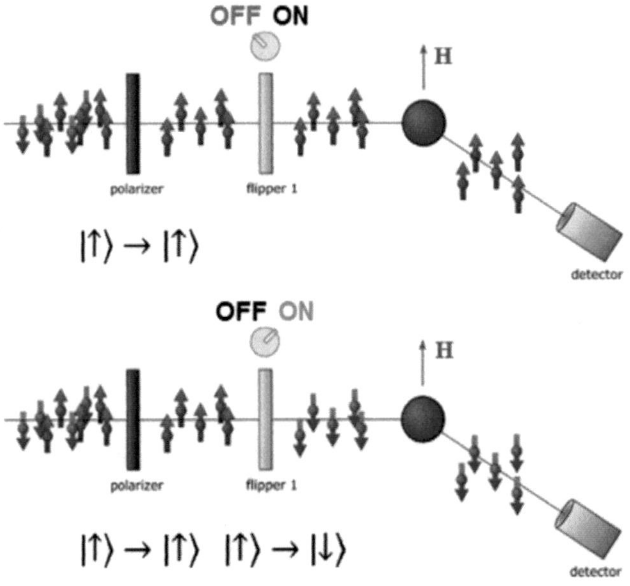

Figure 16. Sketch of a polarized neutron experiment using the flipping ratio method. The spin of the polarized neutrons can be reversed by turning on the spin flipper which is positioned between the monochromator and the sample. Therefore, the intensities I^+ and I^- can be measured in order to derive the flipping ratio $R = I^+/I^-$. An analyzer is not necessary as all scattering should be non-spin-flip.

4.6 Longitudinal polarization analysis

In an LPA experiment the final polarization state of the neutrons needs to be controlled as much as the initial one. Therefore a second polarizer – or rather analyzer – and a second spin flipper are placed between the sample and the detector (Fig. 17).

By setting a magnetic guide field along the desired polarization axis ($\boldsymbol{x} \parallel \boldsymbol{Q}$, $\boldsymbol{y}$ or $\boldsymbol{z}$) at the sample position the neutron spin will adiabatically (due to the superposition of the guide field at the sample and the guide field between the monochromator and the sample) follow the field change. This is effectuated by a set of Helmholtz coils in which different currents are set to produce superposing magnetic fields. Therefore, the Helmholtz coils offer the possibility of

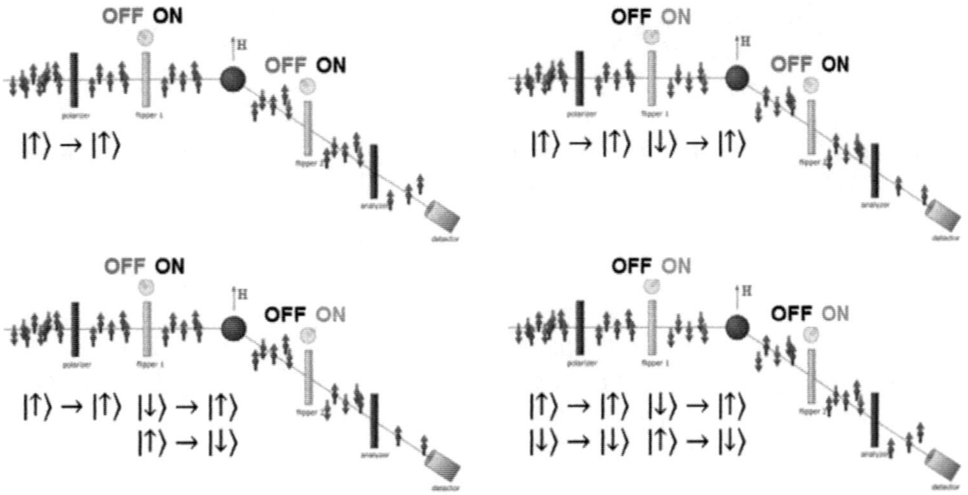

Figure 17. Sketches of a polarized neutron experiment with longitudinal polarization analysis. The combination of a polarizer, an analyzer and two spin flippers makes it possible to measure all four spin-state transitions.

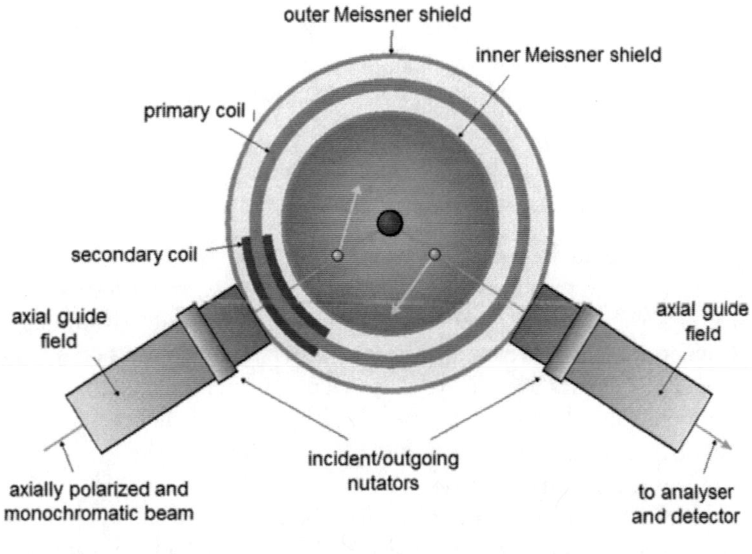

Figure 18. Schematic setup of the CryoPAD.

turning the neutron polarization to one of the three reference axes. The interaction with the sample can be of spin-flip or non-spin-flip nature and in either case the vertical guide field between sample and analyzer will adiabatically turn the polarization towards the z axis so that e.g. only spin-up neutrons can pass the analyzer. Switching on the second flipper inverts the neutron spins and can therefore separate SF and NSF scattering.

4.7 Spherical polarization analysis

Much more information can be gathered when the full polarization matrix of the scattered neutron spin is investigated, i.e. one can entirely trace the change of the neutron spin after the interaction with the sample. As the cross sections for different incident and final polarizations are known important information about the magnetic moment configuration in the studied system can be derived, especially chiral terms. The instrumental realization of such a task is done using the CryoPAD (Fig. 18). In order to have full control of the neutron spin the sample needs to be in an absolutely magnetic field free region which is achieved by an inner Meissner shield provided by a superconducting material. In order to manipulate the neutron spin in the incident and in the scattered beam two individual magnetic fields are necessary which are supplied by a primary coil and a secondary. On the incoming and the outgoing parts of the CryoPAD nutators are positioned in which an axial guide field can be set along a direction perpendicular to the direction of motion of the neutron. The neutron spin will then adiabatically rotate towards that field direction. An outer Meissner shield prevents the superposition of the nutator field and the coils. Due to the abrupt field change at the entrance of the CryoPAD the neutron spin will non-adiabatically rotate around the new field direction. Hereby, the magnetic field is set according to the distance the neutron will travel and its wavelength in order to control the rotation angle in such a way that the neutron spin arrives at the desired polarization axis when it enters the field-free region where it stays constant. Depending on the interaction with the sample the neutron spin might be flipped or rotated. In any case by adjusting the field in the primary coil and in the outgoing nutator the x, y or z component of the scattered neutron spin can be rotated to be parallel to the guide field between the analyzer and the sample. Similar to the LPA experiment a set of two spin flippers (after the monochromator and before the analyzer) is used to distinguish between the different spin-flip and non-spin-flip transitions.

Detailed information is available in the chapter *Spherical Neutron Polarimetry* by P.J. Brown in [5].

5. Summary

Different types of diffractometers have been elucidated throughout this chapter. Understanding the origin of their characteristics allows one to adopt the ideal instrument setup for a particular experimental task. Detailed structure analysis profits from high resolution, while weak magnetic signals are best investigated with a high neutron flux. Unfortunately, a compromise between both has to be made. Due to the statistical orientation of the crystallites in a powder no special care has to be taken for positioning the sample in comparison to single-crystal experiments, where different measuring geometries exist. Finally, it was shown that the use of polarized neutrons gives access to much more information when it comes to the investigation of magnetic materials. The analysis of the neutron spin after the scattering process allows distinguishing between nuclear and magnetic scattering and offers the possibility to draw conclusions concerning the configuration of magnetic moments within the sample.

References

[1] `www.ill.fr`
[2] N. Qureshi et al., Phys. Rev. B **82**, 184521 (2010)
[3] U.W. Arndt, B.T.M. Willis, *Single crystal diffractometry* (Cambridge University Press, 1966)
[4] W.R. Busing, H.A. Levy, Acta Cryst. **22**, 457 (1967)
[5] T. Chatterji, ed., *Neutron scattering from magnetic materials* (Elsevier B. V., 2006)

EPJ Web of Conferences **155**, 00003 (2017)
JDN 22
DOI: 10.1051/epjconf/201715500003

Nuclear structures: Twinning and modulation in crystals

Václav Petříček and Michal Dušek

Institute of Physics, Czech Academy of Sciences v.v.i., Na Slovance 2, 182 21Praha 8, Czech Republic

Abstract. Crystal structure analysis is a standard technique routinely applied to single crystals as well as powders. However the process is not so straightforward if the crystal sample is affected by twinning or if the structure is modulated. In such cases the standard procedures are not directly applicable. The main purpose of this contribution is to show how to solve and refine such difficult structures. While for twinned structures the basic property of crystal – translation symmetry in three dimensional space–remains valid, for modulated crystals a special superspace theory must be exploited in order to describe the atomic structure with crystallographic methods generalized for superspace.

1. Introduction

Modern structure analysis of single crystals based on diffraction experiment is nowadays a standard discipline which allows solving and refining structures of most new crystalline materials. Data are usually acquired on laboratory X-ray diffractometers, and structure solution and refinement is made with standard program packages such SHELX [1], Olex2 [2], Crystals [3] or Jana2006 [4] in almost automatic way. Such a work can be made by a non-specialist in crystallography because most of possible problems can be detected and corrected with help of checking program such as Platon [5]. Simple structure of a well diffracting sample can be now measured, solved and refined in less than one hour and the task can be done without a knowledge what is behind these powerful tools.

However, not all structures can be solved by using these standard approaches. Two effects – twinning and modulations – often make serious problems in structure solution and refinement if they are present in the crystals. These effects are especially important during phase transitions where electric and/or magnetic properties of the studied crystal may depend substantially on the phase and therefore a full description of the corresponding structural changes is very important. If these changes affect atomic positions as well as ordering of magnetic moments, such crystals must be studied by both X-ray and neutron diffraction technique. In this paper we shall concentrate on problems of solution and refinement of nuclear structure from the neutron diffraction experiment. However, most of explanations are also valid for X-ray diffraction.

In spite of the fact that most of standard programs can handle twinning in the crystal, application of these tools needs much deeper understanding of crystallography. Even more complicated are modulated crystals, which miss the basic property of the classical crystal, i.e. translation symmetry. In this paper we will introduce shortly the problem of solution and refinement of difficult structures. For more details about the underlying theory, we recommend the monographs [6] and [7].

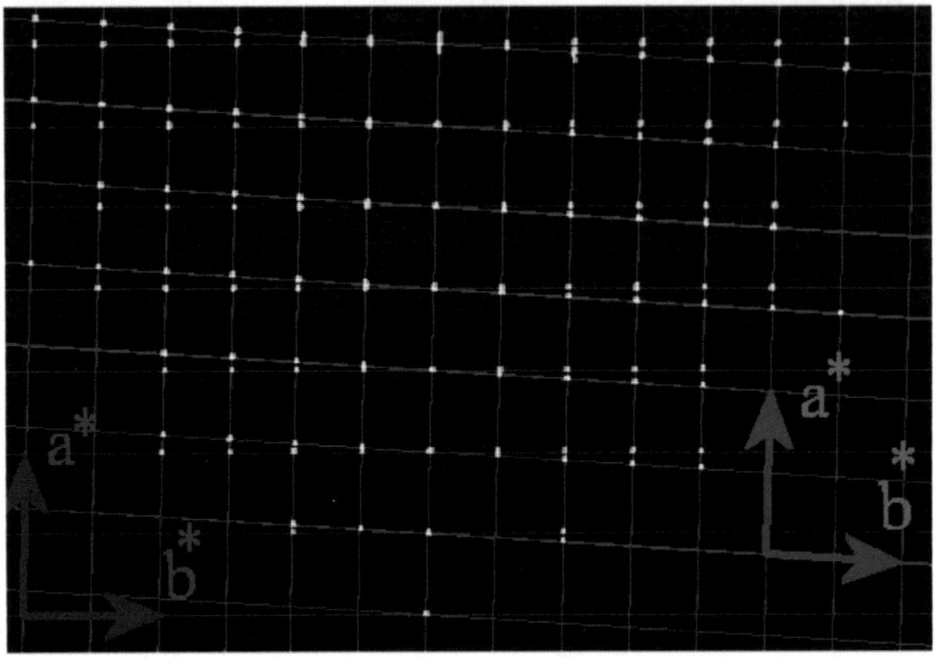

Figure 1. Diffraction pattern of monoclinic crystal, with monoclinic angle β different for 90°, twinned by a general rotation 180° along c axis.

2. Twinning

2.1 Overlaps in twinned crystals

Twinned crystals are supposed to be composed from structurally identical domains mutually related by proper or improper rotations. The number of different domains is usually not very large and in most cases there are present just two differently oriented domains. The simplest case leads to existence of two mutually rotated diffraction patterns which can be simply recognized from diffraction pattern, see Fig. 1.

In this example, reflections *(h,k,0)* are systematically overlapped with reflections $(-h,-k,0)$ of the second domain. Depending on the monoclinic angle β, additional (random) overlapping reflections may occur for (h,k,l) with l being close to $2na/c\cos\beta$.

The relationship between orientations of two domains is related by so called twinning matrix $\mathbf{T}_2$which is defined by the equation:

$$\left[\mathbf{a}_1\ \mathbf{b}_1\ \mathbf{c}_1\right] = \left[\mathbf{a}_2\ \mathbf{b}_2\ \mathbf{c}_2\right] \cdot \mathbf{T}_2 \tag{1}$$

where $\mathbf{a}_i$, $\mathbf{b}_i$ and $\mathbf{c}_i$ are cell parameters of the ith domain. The twinning matrix can be as used to express diffraction indices of the second domain in the reciprocal coordinate system of the first (reference) system:

$$\left[h_{21}\ k_{21}\ l_{21}\right] = \left[h_2\ k_2\ l_2\right] \cdot \mathbf{T}_2 \tag{2}$$

While the indices (h_2,k_2,l_2) are integers defining some reflections from the second domain, their corresponding coordinates (h_{21},k_{21},l_{21}) in the reference subsystem of the first domain need not to be integers. The twinning matrix can be used for predicting the full and partial

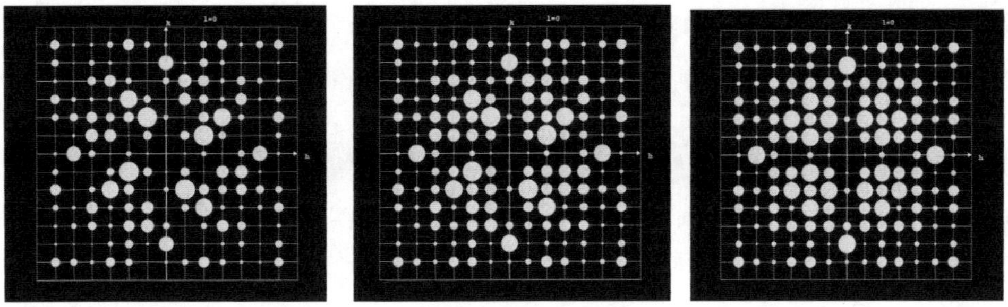

(a) (b) (c)

Figure 2. $(h, k, 0)$ diffraction plane of a simulated structure with point group 4/m: (a) without twinning; (b) twinned by 180° rotation along b axis with volume fractions (3:1); (c) twinned by 180° rotation along b axis with volume fractions (1:1).

overlaps of diffraction spots, using the distance in reciprocal space between (h_{21}, k_{21}, l_{21}) and the closest (h_1, h_1, l_1). The reflections are supposed to be fully overlapped if their distance is smaller than the resolution limit of the data collection and fully separated if the distance is larger than the selected limit for such separation. Reflections having partial overlaps, i.e. with their distance larger than the resolution limit but smaller than the limit for full separation should be discarded from refinement.

Such a criterion is very rough and it might lead to relatively large number of deleted reflections. Moreover, the resolution and separation limits are not the same for all pairs of reflections, because they may be registered in an area detector at different geometry. For this reason, the standard data processing programs like Crysalis, Saint or Eval recognize the overlaps during data processing and encode corresponding information (i.e. which reflections are overlapped and separated) in a form of so-called hklf5 format. In such case the partially overlapped reflections are considered together in the refinement and no information needs to be deleted.

The case presented in the Fig. 1 is simple: the twinning is easily recognized from the diffraction pattern, the structure is solved by standard techniques from one domain or from a detwinned data and, finally, the structure is refined using the twinning matrix or hklf5 format.

Twins with full overlap of all reflections are much more difficult case since they cannot be easily recognized. In the Fig. 2 we present a simulated diffraction pattern of a tetragonal structure with point symmetry 4/m, having different portion of twin domains related by 180° rotation along b axis. In cases (b) and (c) no indication of twinning is apparent and, moreover, the symmetry of the diffraction pattern (c) is higher (*4*/mmm) than the Laue symmetry of the structure. This means that presence of twinning in the crystal may obscure the correct symmetry, which can be deduced only during solution and refinement process. From Fig. 2, the necessary condition for such full overlaps is obvious: the point symmetry of the reciprocal lattice must be higher than the point symmetry of the structure.

Similar effects can be present even for structures of lower symmetry (triclinic, monoclinic, ...) if their cell parameters correspond within the experimental accuracy to a higher lattice point symmetry. This may happen as a result phase transitions in which the original symmetry is reduced but cell parameters are almost unchanged. In the following we shall concentrate on problems of data processing, solution and refinement of twins with full overlaps.

2.2 Symmetry of the twinned diffraction pattern

Overlapping of diffraction spots from n twin domains can be expressed by the following formula for the combined structure factor $\mathcal{F}(\mathbf{h})$:

$$\mathcal{F}^2(\mathbf{h}) = v_1 F^2(\mathbf{h}\mathbf{T}_1) + v_2 F^2(\mathbf{h}\mathbf{T}_2) + \ldots + v_n F^2(\mathbf{h}\mathbf{T}_n) \tag{3}$$

where v_i is the volume fraction of ith domain and $F(\mathbf{h})$ is the structure factor. The volume fractions are normalized to 1:

$$\sum_{i=1}^{n} v_i = 1 \tag{4}$$

The Eq. (3) is based on the assumption that diffractions from twin domains are independent. The structure factors $F(\mathbf{h})$ are coefficients in the Fourier summation of periodic density of scatterers $\rho(\mathbf{r})$:

$$\rho(\mathbf{r}) = \rho\left(\mathbf{r} + \sum_{i=1}^{3} n_i \mathbf{a}_i\right) \tag{5}$$

$$\rho(\mathbf{r}) = \sum_{\mathbf{h}} F(\mathbf{h}) exp(-2\pi i \mathbf{h}\mathbf{r}) \tag{6}$$

The combined structure factor is $\mathcal{F}(\mathbf{h})$ and it can be calculated from the integrated intensity $I(\mathbf{h})$ of the relevant diffraction spot as for a non-twinned crystal:

$$I(\mathbf{h}) = S.Lp(|\mathbf{h}|).A(\mathbf{h}).\mathcal{F}^2(\mathbf{h}) \tag{7}$$

where S stands for the scaling factor, A for absorption correction and Lp for Lorentz-polarization factor.

The full symmetry of the crystal is expressed by the space group symmetry which consists from the symmetry operations of the type:

$$\mathbf{r}' = \mathbf{R}_j \mathbf{r} + \mathbf{s}_j + \mathbf{c}_k + \mathbf{n} \tag{8}$$

where $\mathbf{R}_i$ and $\mathbf{s}_i$ are rotation and translation part of the ith symmetry operation, $\mathbf{c}_j$ is j^{th} centering vector and $\mathbf{n}$ is an arbitrary unit cell translation. Then the generalized equation for symmetry condition is:

$$\rho(\mathbf{r}) = \rho\left(\mathbf{R}_j \mathbf{r} + \sum_{i=1}^{3} (s_{ji} + c_{ki} + n_i)\mathbf{a}_i\right) \tag{9}$$

From the Eqs. (9) and (6) we can get:

$$F(\mathbf{h}\mathbf{R}_j) = F(\mathbf{h}) exp[2\pi i \mathbf{h}(\mathbf{s}_j + \mathbf{c}_k)] \tag{10}$$

and for squared structure factors the following symmetry relationship holds:

$$F^2(\mathbf{h}\mathbf{R}_j) = F^2(\mathbf{h}) \tag{11}$$

This means that the diffraction pattern follows the point group symmetry. Moreover, the fact that the nuclear density is a real function leads to the following relationship for the complex conjugated structure factor:

$$F^*(\mathbf{h}) = F(-\mathbf{h}) \tag{12}$$

and therefore the diffraction pattern shows contains always an inversion center even for non-centrosymmetric structures.

As mentioned above, complete overlaps of diffraction spots of the twinned crystals are possible only if the point symmetry of the lattice $\mathcal{H}$ is higher than the point symmetry of the structure $\mathcal{G}$. Say that the order of the subgroup $\mathcal{G}$ with respect to $\mathcal{H}$ is n. Then the point group $\mathcal{H}$ can be decomposed into left cosets:

$$\mathcal{H} = \mathbf{T}_1\mathcal{G} + \mathbf{T}_2\mathcal{G} + \ldots + \mathbf{T}_n\mathcal{G} \tag{13}$$

where the set of twinning operations $\{\mathbf{T}_1, \mathbf{T}_2, \ldots .\mathbf{T}_n\}$ are selected arbitrary as representatives of each coset. Without loss of generality the first twinning element can be chosen as an identity operation.

Symmetry of the diffraction pattern in the case of equal volume fractions follows the lattice symmetry $\mathcal{H}$. But as volume fractions need not be exactly equal we cannot a priory use during data merging all symmetry operations from $\mathcal{H}$. Question is if all symmetry operations from the point group $\mathcal{G}$ are always present in the diffraction pattern. For the reflection $\mathbf{hG}_j$ where $\mathbf{G}_j$ is an arbitrary operation from $\mathcal{G}$ the Eq. (3) gives:

$$\mathcal{F}^2(\mathbf{hG}_j) = v_1 F^2(\mathbf{hG}_j\mathbf{T}_1) + v_2 F^2(\mathbf{hG}_j\mathbf{T}_2) + \ldots + v_n F^2(\mathbf{hG}_j\mathbf{T}_n)$$

then the symmetry requirement $\mathcal{F}^2(\mathbf{hG}_j) = \mathcal{F}^2(\mathbf{h})$ leads to set of conditions:

$$\mathbf{G}_j\mathbf{T}_k = \mathbf{T}_k\mathbf{G}_l \Rightarrow \mathbf{G}_l = \mathbf{T}_k^{-1}\mathbf{G}_j\mathbf{T}_k \tag{14}$$

where $\mathbf{G}_l$ is again an operation from $\mathcal{G}$. These conditions are fulfilled for a subgroup $\mathcal{G}$ only if it is a normal subgroup in $\mathcal{H}$. In case the subgroup $\mathcal{G}$ is not a normal subgroup of H only operations fulfilling the Eq. (14) can be used in the merging procedure of symmetry equivalent reflections. Note that if the order of the subgroup $\mathcal{G}$ in $\mathcal{H}$ is two, the subgroup is always normal.

The conclusion about symmetry of the diffraction pattern is that the symmetry can mimic the lattice symmetry but it can even mimic a symmetry lower than the point group of the crystal.

2.3 Systematic extinctions of twins

Systematic extinctions are used to determine a correct space group of the studied compound. From the Eq. (10) it follows that reflections fulfilling the condition $\mathbf{hR}_j = \mathbf{h}$ (i.e. $\mathbf{h}$ invariant with respect to $\mathbf{R}_j$) can have non-zero structure factor only if:

$$\exp\left[2\pi i\mathbf{h}\left(\mathbf{s}_j + \mathbf{c}_k\right)\right] = 1 \quad \Rightarrow \quad \mathbf{h}\left(\mathbf{s}_j + \mathbf{c}_k\right) = n \tag{15}$$

where n is an arbitrary integer.

Example: The mirror plane with its normal parallel to the **c** axis has the following matrix form with respect to the crystal coordinate system **a**, **b** and **c**:

$$\left(\begin{bmatrix} 1 & 0 & 0 \\ 0 & 1 & 0 \\ 0 & 0 & -1 \end{bmatrix}, \begin{bmatrix} s_1 \\ s_2 \\ s_3 \end{bmatrix}\right) \tag{16}$$

From the space group theory it follows that in a primitive unit cell s_1 and s_2 are equal either to 0 or $1/2$ and that the third component s_3 in not restricted and depends on the origin selection. Taking into account all combinations and the Eq. (15), we can get extinctions as summarized in the Table 1.

Table 1. Systematic extinctions for $m(001)$.

(s_1, s_2, s_3)	**Symmetry operation**	**Extinction condition**
$(0, 0, 0)$	m	–
$(1/2, 0, 0)$	a	$h = 2n + 1$
$(0, 1/2, 0)$	b	$k = 2n + 1$
$(1/2, 1/2, 0)$	n	$h + k = 2n + 1$

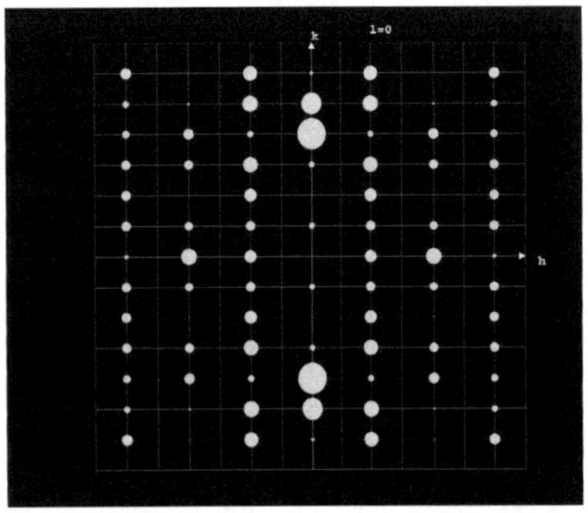

(a)

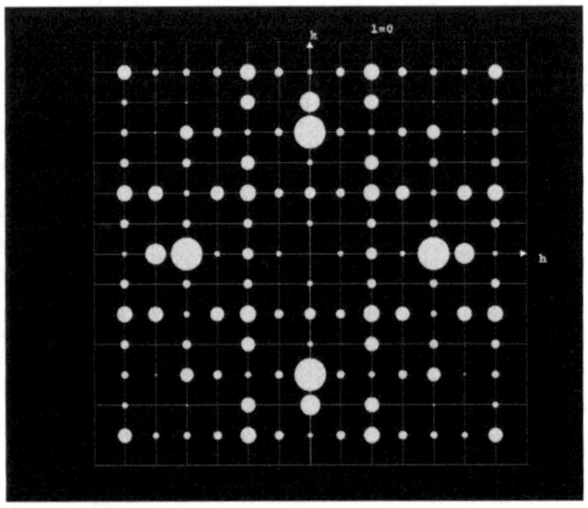

(b)

Figure 3. $(h, k, 0)$ diffraction plane of simulated structure with an a-glide plane: (a) without twinning; (b) twinned by 90° rotation around c axis with volume fractions 1:1.

Overlapping of reflections in twinned crystals can partially hide the extinction conditions. In the example shown in the Fig. 3 such an effect is demonstrated for the a-glide plane:

As a result of the twinning, the absent reflections must fulfill the conditions $h = 2n + 1$ and $k = 2n + 1$, which cannot occur for a non-twinned crystal. Thus the extinctions are violated in very specific way which can help to recognize that the crystal in twinned.

2.4 Solution and refinement of twinned structures

Solution of crystal structures of twinned samples with completely overlapping diffraction patterns may be a difficult task. In cases where twinning is a result of a phase transition from higher to lower symmetrical phase, the known structural model of the higher symmetrical phase can be used as a starting point for the structural solution of the lower symmetrical phase. However, in many cases several different subgroups are to be tested to find correct solution.

In the case that domains are not equally occupied and an approximate volume fraction is known, the observed structure factors can be corrected to get "detwinned" structure factors useful for ab-initio solution by standard methods:

$$F^2(\mathbf{h}) = [(1 - v)\mathcal{F}^2(\mathbf{h}) - v\mathcal{F}^2(\mathbf{h} \cdot \mathbf{T}_2)]/(1 - 2v) \quad (17)$$

However, this correction cannot be used for equally or almost equally occupied domains.

On the other hand, as soon as some starting structural model is known, the refinement and completing the structure is very similar to that of a regular structure. The number of parameters in the refinement is enlarged only by the (n-1) volume fractions.

For completing the structure a Fourier maps can be used in which the observed structure factors are corrected for twinning by one of these methods:

$$F^2_{obs,corr}(\mathbf{h}) = \frac{1}{v_1}\left[\mathcal{F}^2_{obs}(\mathbf{h}) - v_2 F^2_{calc}(\mathbf{h}\cdot\mathbf{T}_2) - v_3 F^2_{calc}(\mathbf{h}\cdot\mathbf{T}_3) - \ldots - v_n F^2_{calc}(\mathbf{h}\cdot\mathbf{T}_n)\right] \tag{18}$$

$$\left|F_{obs,corr}(\mathbf{h})\right| = \left|\mathcal{F}_{obs}(\mathbf{h})\right| \frac{|F_{calc}(\mathbf{h})|}{|\mathcal{F}_{calc}(\mathbf{h})|}$$

The above described methods handling twinned crystals have been implemented in the program Jana2006 and several typical practical examples were included into Jana cookbook (examples 3.1–3.4). Both can be downloaded the Jana home page: `http://jana.fzu.cz/`.

3. Modulated crystals

3.1 Superspace approach for modulated structures

Translational periodicity is the basic property of the classical crystal, see the Eq. (5). A direct consequence of this periodicity is the existence of a diffraction pattern by X-ray, neutrons or electrons with sharp diffraction spots which are localized at the points of the reciprocal lattice defined by basic vectors $\mathbf{a}_1^*$, $\mathbf{a}_2^*$ and $\mathbf{a}_3^*$:

$$\mathbf{h} = h_1\mathbf{a}_1^* + h_2\mathbf{a}_2^* + h_3\mathbf{a}_3^* \tag{19}$$

The reciprocal lattice vectors are related to translational (direct space) lattice vectors by equations:

$$\mathbf{a}_1^* = \frac{\mathbf{a}_2 \times \mathbf{a}_3}{\mathbf{a}_1(\mathbf{a}_2 \times \mathbf{a}_3)} \quad \mathbf{a}_2^* = \frac{\mathbf{a}_3 \times \mathbf{a}_1}{\mathbf{a}_1(\mathbf{a}_2 \times \mathbf{a}_3)} \quad \mathbf{a}_3^* = \frac{\mathbf{a}_1 \times \mathbf{a}_2}{\mathbf{a}_1(\mathbf{a}_2 \times \mathbf{a}_3)} \tag{20}$$

However, there are compounds which also give additional sharp diffraction spots, so-called satellites, localized out of the nodes of the reciprocal lattice. These spots can be indexed only if one or more additional (so called modulation) vectors are used:

$$\mathbf{H} = h_1\mathbf{a}_1^* + h_2\mathbf{a}_2^* + h_3\mathbf{a}_3^* + m\mathbf{q}\ldots \tag{21}$$

In most cases one modulation vector is sufficient for indexing of satellites, and in the following text we shall confine ourselves to the case with one modulation vector. The diffraction pattern of such modulated crystal looks like the one shown in the Fig. 4.

The fact that the additional (satellite) reflections are sharp and regularly distributed in the diffraction space means that the violation of the 3d translation symmetry must be somehow regular. For handling of modulated structure a special theory of superspace and its symmetry has been developed. The development is closely connected with names Peter de Wolff, Aloysio Janner and Ted Janssen [8].

The superspace approach is based on the construction which moves artificially all satellite reflections into the 4th dimension with a shift proportional to the satellite index – see Fig. 5. The spots are now making a lattice in the four dimensional reciprocal space described by the reciprocal vectors:

$$\mathbf{A}_i^* = \mathbf{a}_i^* \quad \text{for } i = 1, 2, 3 \quad \mathbf{A}_4^* = \mathbf{q} + \mathbf{e} \tag{22}$$

The lattice character of the diffraction pattern in $(3+1)$d superspace means that the modulated structure has a translation periodicity in $(3+1)$d superspace. Thus the generalized

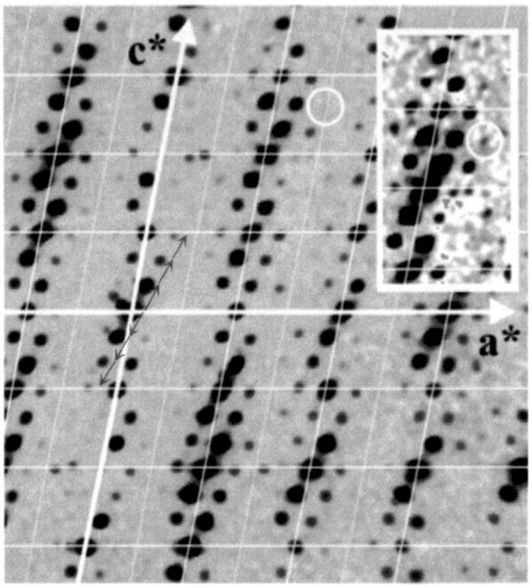

Figure 4. (*h,0,l*) diffraction plane of Na_2CO_3. The white grid represents the lattice of the main reflections. The additional (satellite) diffraction spots are regularly displaced from regular (main) spots by a q-vector (red arrows).

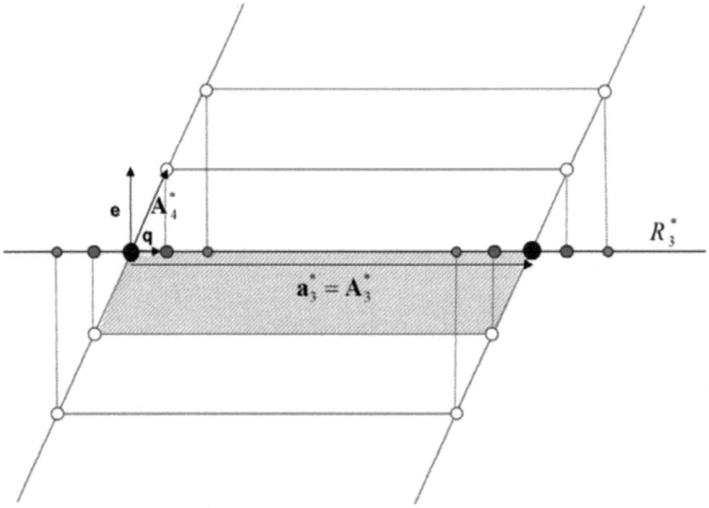

Figure 5. Introducing superspace in the reciprocal space: the vector **e** is perpendicular to the diffraction space R_3^*. Black spots are main reflections located at the nodes of the three-dimensional reciprocal lattice. Gray spots are satellite reflections. The new reciprocal lattice vector: $\mathbf{A}_4^* = \mathbf{q} + \mathbf{e}$. White spots are satellites projected to the new lattice vector; the blue area is a $(3+1)$d reciprocal cell describing main reflections as well as satellites.

nuclear density function fulfills the equation:

$$\tilde{\rho}(\mathbf{r}) = \tilde{\rho}\left(\mathbf{r} + \sum_{i=1}^{4} n_i \mathbf{A}_i\right) \tag{23}$$

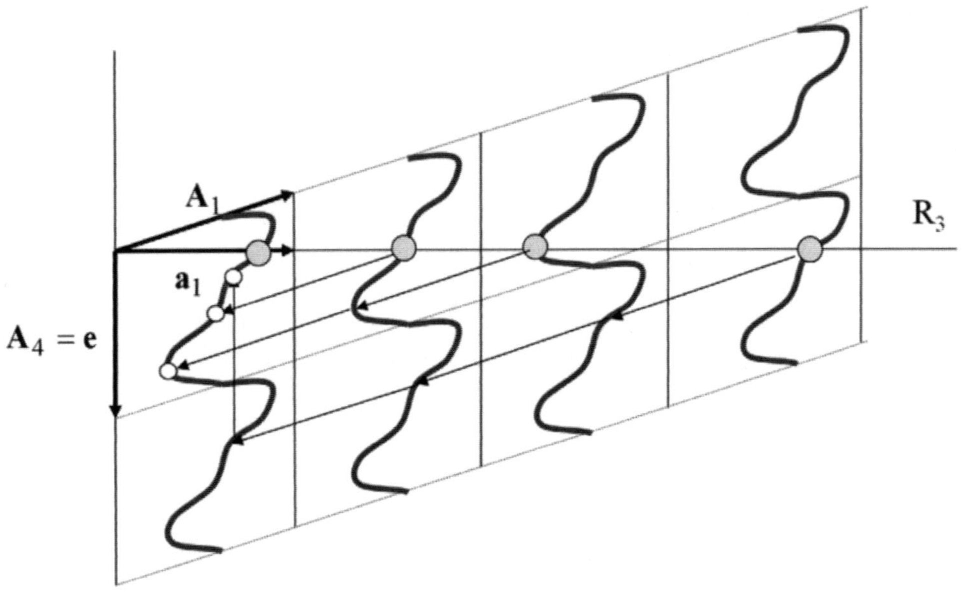

Figure 6. The positionally modulated structure in the (3 + 1)d superspace: the atomic positions are described by the modulation function defined in a (3 + 1)d unit cell. This unit cell can be repeated by the (3 + 1)d translational symmetry to fill the whole (3 + 1)d superspace. Large blue points indicate intersections of the modulation function with R_3 (the real space). These intersections define the crystal structure. Due to the translation symmetry, each such point can be transformed to the first unit cell (as white points).

where the lattice vectors $\mathbf{A}_i$ fulfil the equations:

$$\mathbf{A}_i = \mathbf{a}_i - (\mathbf{a}_i \cdot \mathbf{q})\ \mathbf{e} \qquad i = 1, 2, 3 \quad \mathbf{A}_4 = \mathbf{e} \tag{24}$$

as follows from orthogonality conditions between direct $\mathbf{A}_i$ and reciprocal $\mathbf{A}_i^*$:

$$\mathbf{A}_i \cdot \mathbf{A}_j^* = \delta_{ij}$$

In the Fig. 6 the translation symmetry is demonstrated for positionally modulated structure. There is again a unit cell which contains structural information needed to generate whole modulated structure. The superspace approach allows generalization of standard structure tools such as Fourier and Patterson syntheses, structure solution and refinement techniques.

As the really observed 3d diffraction pattern is a projection from an auxilliary (3 + 1)d diffraction pattern in superspace along the vector **e**, corresponding 3d Fourier map which shows a nuclear density, is a section perpendicular to the vector **e** through a (3 + 1)d periodic superspace map. Moving this section along the vector **e** gives sections (so-called t-sections) of the same modulated structure which differs only by the origin shift.

The (3 + 1)d Fourier maps are calculated from structure factors according to the equation analogical to the Eq. (6). Structure factor amplitudes are based on observed intensities (Eq. (7)) while phases follow from the actual model. The sections, in which modulation of one atomic coordinate x_1, x_2 or x_3 is visualized as a function of x_4, are called de Wolff's sections. They play a crucial role in finding and completing modulation models for individual atoms in the structure.

3.2 Superspace symmetry

Symmetry operation $\hat{S}$ in the superspace must keep the generalized electron density invariant:

$$\tilde{\rho}(\hat{S}\mathbf{r}) = \tilde{\rho}(\mathbf{r}) \tag{25}$$

A set of symmetry operations fulfilling this equation constitutes a superspace group. In analogy to 3d structures, the superspace symmetry induces diffraction symmetry, and the whole diffraction pattern including satellite reflections follows the corresponding point group. Introducing symmetry in $(3+1)$d superspace does not mean that all possible four-dimensional space groups are available for crystal symmetry. There are some restrictions which considerably reduce the number of four dimensional space groups acceptable for description of modulated structures. Any superspace operation $\mathbf{R}$ has the following matrix form with respect to the given crystallographic base:

$$\mathbf{R} = \left(\left[\begin{array}{c|c} \mathbf{\Gamma}_E & \mathbf{0} \\ \hline \mathbf{\Gamma}_M & \mathbf{\Gamma}_I \end{array} \right], \left[\begin{array}{c} \mathbf{s}_E \\ \hline \mathbf{s}_I \end{array} \right] \right) \tag{26}$$

The left-upper part $\mathbf{\Gamma}_E$ is 3×3 matrix which represents proper or improper rotation in three dimensional space (known from 3d space groups). The right-upper part is $3\times$ 1 matrix (a column) of zeros. This restriction follows from the fact that superspace is based in a mathematical construction which uses an auxilliary vector $\mathbf{e}$. This vector must stay perpendicular to the real three dimensional space when the symmetry operation is applied. The right-lower block $\mathbf{\Gamma}_I$ is 1×1 matrix with value equal ± 1. Finally, the left-lower block $\mathbf{\Gamma}_M$ is 1×3 matrix (row) fulfilling the equation:

$$\mathbf{q}\mathbf{\Gamma}_E - \mathbf{\Gamma}_I\mathbf{q} = \mathbf{\Gamma}_M \tag{27}$$

This means that the rotation part of the symmetry operation is fully determined from the 3d rotation part and modulation vector.

The translation part $\mathbf{s}_E$ follows from the corresponding 3d space group. The only new (compared with standard crystallography) part is $\mathbf{s}_I$ which defines the translation component along the additional 4th direction. Value of this component can induce systematic extinctions of satellite reflections. For example the mirror plane with its normal parallel to the $\mathbf{c}$ axis and $\Gamma_I = 1$ causes extinctions as described in the Table 2. In this table, the symbol used for symmetry elements consists from two parts, the first one defines the 3d symbol and the second one the additional information about $\mathbf{s}_I$, which in our example is 0 and s for $\mathbf{s}_I$ equal 0 and 1/2, respectively.

The superspace symbols are described in the International Tables for Crystallography (IT) vol. C [9]. The symbol consists of three parts – space group symbol, specification of the modulation vector and specification of the $\mathbf{s}_I$ components of the symmetry operations. The components of modulation vector are either restricted by the Eq. (27) to specific rational values or they can have general (irrational) value. Irrational components are specified by Greek letters α, β or γ. Example: $Pmna(0\ {}^1\!/\!_2\ \gamma)s00$. Information available in the IT vol. C is limited, without a list of superspace group operations which restrict modulation functions of atoms located at special positions. Fortunately the list of symmetry operations for superspace groups up to $(3+3)$d superspace as well as additional details about their standard settings is accessible from [10]. The symmetrical restrictions of modulated functions expressed as a combination of harmonic waves for $(3+1)$ superspace groups have been published [11] and they are automatically applied in the program Jana2006.

Table 2. Systematic extinctions for $m(001)$.

(s_1, s_2, s_3, s_4)	Symmetry operation	Extinction condition
$(0, 0, 0, 0)$	$\binom{m}{0}$	–
$(0, 0, 0, 1/2)$	$\binom{m}{s}$	$m = 2n + 1$
$(1/2, 0, 0, 0)$	$\binom{a}{0}$	$h = 2n + 1$
$(1/2, 0, 0, 1/2)$	$\binom{a}{s}$	$h + m = 2n + 1$
$(0, 1/2, 0, 0)$	$\binom{b}{0}$	$k = 2n + 1$
$(0, 1/2, 0, 1/2)$	$\binom{b}{s}$	$k + m = 2n + 1$
$(1/2, 1/2, 0, 0)$	$\binom{n}{0}$	$h + k = 2n + 1$
$(1/2, 1/2, 0, 1/2)$	$\binom{n}{s}$	$h + k + m = 2n + 1$

3.3 Basic modulation types

Modulations in the crystal can affect all structural parameters i.e. atomic occupancies, positions, ADP, multipole coefficients or magnetic moments. Any structural parameter $p(x_4)$ due to the translation symmetry in the superspace can be expressed as a Fourier expansion:

$$p(x_4) = p_0 + \sum_n \left(p_{s,n} \sin 2\pi n x_4 + p_{c,n} \cos 2\pi n x_4\right) \tag{28}$$

where p_0, $p_{s,n}$, $p_{c,n}$ are expansion coefficients which are used to describe modulation wave. The number of used terms in the Eq. (28) depends on complexity of the modulation in crystal which is usually connected with the maximal number of observable satellite reflections. In the following a simple modulation models and their influence to the diffraction pattern is presented.

3.3.1 Occupational modulation

The simplest occupational modulation with only one harmonic wave can be expressed like:

$$p(x_4) = p_0 + p_{s,1} \sin(2\pi x_4) + p_{c,1} \cos(2\pi x_4) \tag{29}$$

From the kinematical theory of diffraction it can be shown that such modulation will generate only first order satellites, as visualized in the Fig. 7(a). The corresponding de Wolff's section through the (3 + 1)d Fourier map is shown in Fig. 7(b):

This simple harmonic model for occupation modulation is not very common in real-life examples. Usually the diffraction pattern also contains higher order satellites and the description of modulations needs more harmonic waves. In the limiting case the modulation takes step like character:

$$\begin{aligned} p(x_4) &= 1 \qquad x_4 \in \langle x_4^0 - \tfrac{\Delta}{2}, x_4^0 + \tfrac{\Delta}{2}\rangle \\ p(x_4) &= 0 \qquad x_4 \notin \langle x_4^0 - \tfrac{\Delta}{2}, x_4^0 + \tfrac{\Delta}{2}\rangle \end{aligned} \tag{30}$$

where x_4^0 and Δ define the center and length of the x_4 interval in which the atom is present.

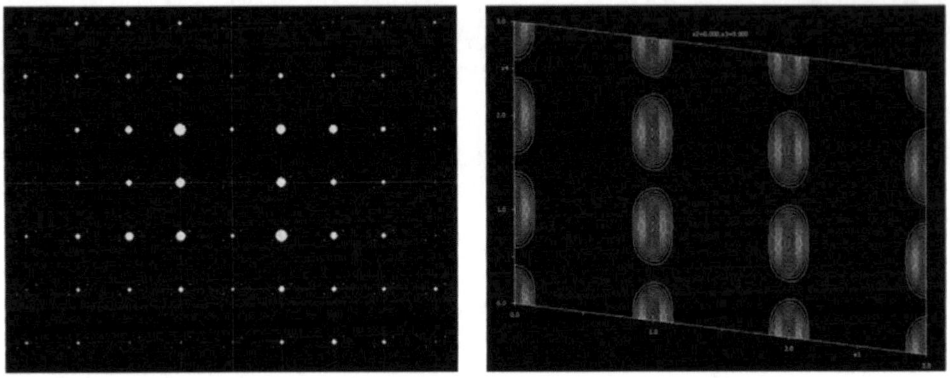

(a) (b)

Figure 7. The simple harmonic occupational modulation: (a) diffraction pattern, (b) de Wolff's section.

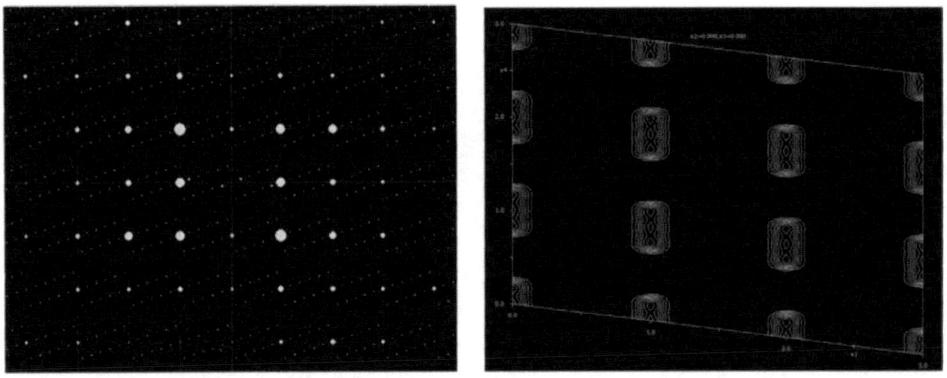

(a) (b)

Figure 8. The step like occupational modulation (so-called crenel modulation): (a) diffraction pattern, (b) de Wolff's section.

As shown in the Fig. 8, the step like modulation leads to a high number of satellite reflections. Such a modulation is called crenel modulation [12]. Description of step-like modulation by harmonic waves would require high number of parameters. Instead of them we can use a crenel function (30) which has only two parameters – the center and the width.

3.3.2 Positional modulation

The diffraction patterns for weak and strong modulation amplitude are shown in the Figs. 9 and 10. On contrary to occupational modulation here one harmonic can generate higher satellites. With growing amplitude of the modulation the order and relative intensity of satellites increases.

3.3.3 Composite structures

These structures consist of two or more modulated subsystems each having its $(3+1)$d reciprocal lattice $\mathbf{A}_i^{*\nu}$ $i = 1, 2, 3, 4$ (ν denotes the subsystem) related to a reference lattice

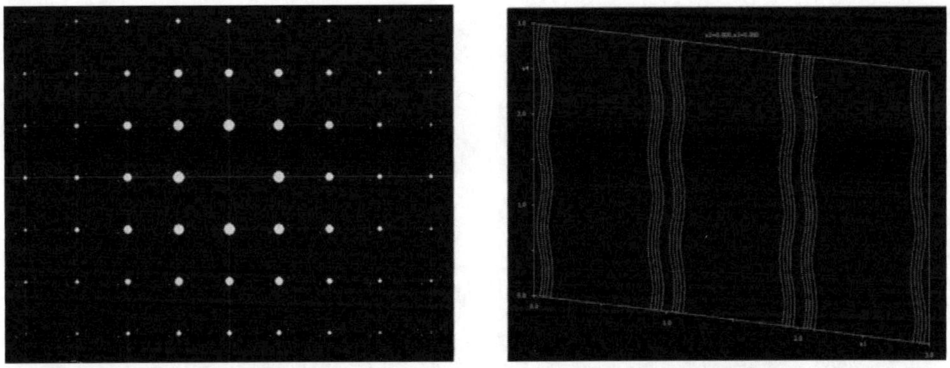

(a) (b)

Figure 9. The weak positional modulation: (a) diffraction pattern, (b) de Wolff's section.

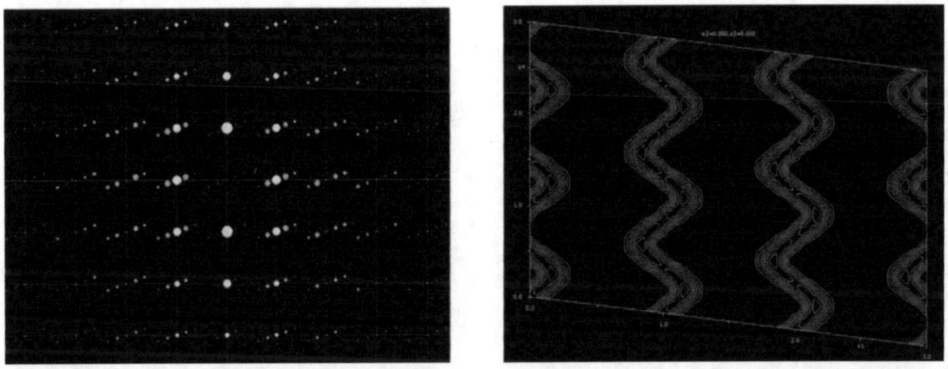

(a) (b)

Figure 10. The strong positional modulation : (a) diffraction pattern, (b) de Wolff's section.

$\mathbf{A}_i^*$ $i = 1, 2, 3, 4$ by the equation:

$$\mathbf{A}_i^{*\nu} = \sum_{i=1}^{4} Z_{ij}^{\nu} \mathbf{A}_j^* \tag{31}$$

where the Z_{ij}^{ν} is an integer matrix. Figure 11 shows a simple example of a composite structure made of two columns of atoms with different periodicity along the $\mathbf{a}_1^1 || \mathbf{a}_1^2$ direction, while the either parameters are identical: $\mathbf{a}_2^1 = \mathbf{a}_2^2$ and $\mathbf{a}_3^1 = \mathbf{a}_3^2$. In this example, the subsystems are not modulated and each of them makes a diffraction pattern which can be fully indexed by three indexes. We need four indexes, however, to index these patterns together.

If we choose the first subsystem as a reference subsystem, the Z matrices take form:

$$\mathbf{Z}^1 = \begin{bmatrix} 1 & 0 & 0 & 0 \\ 0 & 1 & 0 & 0 \\ 0 & 0 & 1 & 0 \\ 0 & 0 & 0 & 1 \end{bmatrix} \quad \mathbf{Z}^2 = \begin{bmatrix} 0 & 0 & 0 & 1 \\ 0 & 1 & 0 & 0 \\ 0 & 0 & 1 & 0 \\ 1 & 0 & 0 & 0 \end{bmatrix}$$

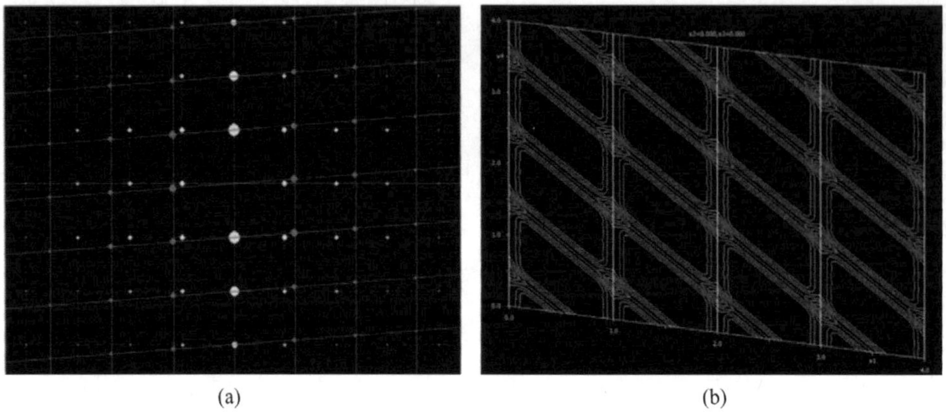

Figure 11. Composite structure composed of non-modulated subsystems: (a) diffraction pattern, (b) de Wolff's section.

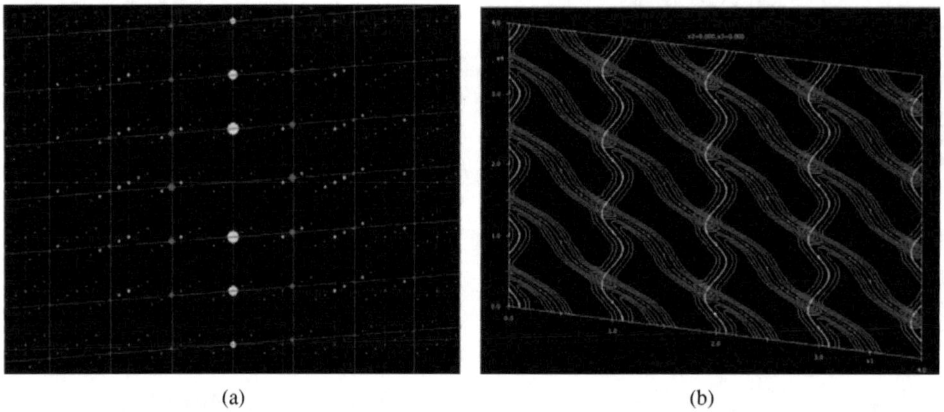

Figure 12. Composite structure composed of modulated subsystems: (a) diffraction pattern, (b) de Wolff's section.

Diffraction pattern of non-modulated composite crystal contains only main reflections of both subsystems. However, the interaction between both subsystems usually induces a mutual modulations as shown in the Fig. 12.

3.4 Solution of modulated structures

In many cases modulated structures can be solved in two steps. In the first step only main reflections are used to solve an average structure by standard methods. Such a structural model can show features like split atomic positions or unusually large ADP ellipsoids (see Fig. 13).

Such effects can help to predict type of modulations and recognize atoms which are strongly modulated. In the second step, the modulation waves are found by refinement from the small randomly chosen displacements.

However in case of strong modulations the second step or even the first step can fail because both the average structure and the initial modulation wave may be too far of the

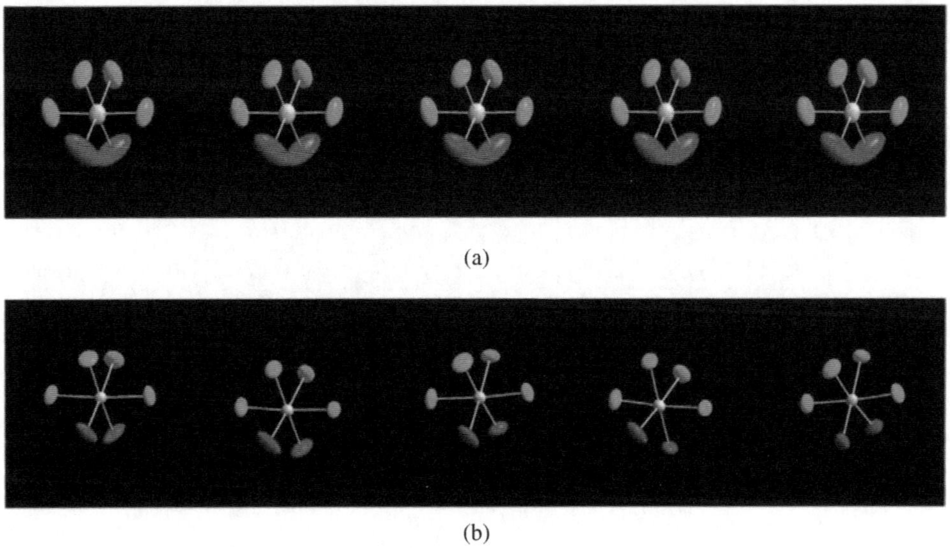

Figure 13. Anion $AsF_4(OH)_2$ [16] (a) average structure calculated from main reflections; (b) modulated structure.

correct solution. As for the standard crystallography, refinement and Fourier methods only work for "almost finished" structures. This was the reason why the standard solution methods were generalized for modulated structures with the aim to solve them *ab initio*. The direct methods based on the Sayre equation were used for the development the program DIMS written by Fan Hai-fu [13, 14]. The heavy atom method for modulated crystals was developed by Steurer [15] and Petricek et al. [16]. However, these methods are not applicable to all cases.

The most promising method for *ab initio* solution of strongly modulated structures is so-called charge flipping [17, 18]. The method has been implemented into the program Superflip [19] and it is distributed together with program Jana2006. This method can solve modulated structures just in one step and starting modulation functions can be deduced directly from the output of the charge flipping, which is the nuclear density map in superspace.

With help of the modern diffractometer software and charge flipping, the steps like indexing, data collection, data processing and structure solution can be done almost routinely. However, it does not mean that complete solution of modulated structures is a routine job. Difficulties start after the solution step, where e.g. interpretation of the charge flipping output requires experience, manual work and analysis of Fourier sections. This interpretation is crucial for proper selection of modulation parameters and it can be specific for different modulated structures. This selection is important not only for refinement but especially for interpretation of results where we aim to understand the reasons why the structure is modulated.

Similarly as for twinned structures, the Jana cookbook contains several worked examples (5.1–5.5 and 7.1–7.3) covering typical cases of modulated structures. Moreover, everybody can participate in workshops which are offered regularly in the Jana web page for people interested in practical solution of modulated structures, as well as in bi-annual workshops organized by university in Bayreuth and in many other events focused to aperiodic crystallography.

References

[1] G. Sheldrick, Acta Cryst. **A64**, 112 (2008)
[2] O.V. Dolomanov, L.J. Bourhis, R.J. Gildea, J.A.K. Howard, H. Puschmann, J. Appl. Cryst. **42**, 339 (2009)
[3] P.W. Betteridge, J.R. Carruthers, R.I. Cooper, K. Prout, D.J. Watkin, J. Appl. Cryst. **36**, 1487 (2003)
[4] V. Petříček, M. Dušek, L. Palatinus, Z. Kristallogr. **229**, 345 (2014)
[5] A.L. Spek, PLATON, *A Multipurpose Crystallographic Tool*, Utrecht University, Utrecht, The Netherlands (2008)
[6] S. van Smaalen, Incommensurate Crystallography. (Oxford University Press, Oxford, 2007)
[7] T. Janssen, G. Chapuis, M. de Boissieu, *Aperiodic Crystals*. (Oxford University Press, Oxford, 2007)
[8] P.M. de Wolff, T. Janssen T., A. Janner, Acta Cryst. **A37**, 625 (1981)
[9] International Tables for Crystallography, Volume C, Ed. E. Prince (Kluwer Acad. Publ., Dordrecht, 2004)
[10] H.T. Stokes, B.J. Campbell, S. van Smaalen, Acta Cryst. **A67**, 45 (2011)
[11] V. Petříček, P. Coppens, Acta Cryst. **A44**, 1051 (1988)
[12] V. Petříček, A. van der Lee, M. Evain, Acta Cryst. **A51**, 529 (1995)
[13] Hao Quan, Liu Yi-wei, Fan Hai-fu, Acta Cryst. **A43**, 820 (1987)
[14] Fan Hai-fu, S. van Smaalen, E.J.W. Lam, P.T. Beurskens, Acta Cryst. **A49**, 704 (1993)
[15] W. Steurer, Acta Cryst. **A43**, 36 (1987)
[16] J. Peterková, M. Dušek V. Petříček, J. Loub, Acta Cryst. **B54**, 809 (1998)
[17] G. Oszlányi, A. Süto, Acta Cryst. **A60**, 134 (2004)
[18] L. Palatinus, Acta Cryst. **A60**, 604 (2004)
[19] L. Palatinus and G.Chapuis, J. Appl. Cryst. **40**, 786 (2007)

Crystallography and dynamics in superspace

Céline Mariette, Philippe Rabiller, Laurent Guérin, and Bertrand Toudic

Institute of Physics Rennes, U.M.R. CNRS 6251, University of Rennes 1, 91944 Rennes, France

Abstract. The discovery in materials of long range order without lattice periodicity has generated a very broad field of research with the introduction of totally new concepts. The common feature of this new state of matter is the translational symmetry recovered by adding supplementary dimensions to the physical space, defining a crystallographic superspace. Along these supplementary directions, specific low excitations, so-called phasons, are predicted and observed. These quasi-periodic crystals may present phase transitions as function of external parameters, which have then to be described as symmetry breakings within these crystallographic superspaces. Three different families of quasi-periodic crystals are usually considered: incommensurate modulated crystals, aperiodic composites and quasicrystals. Their crystallography and dynamics are discussed.

1. Introduction

In the last part of the XXth century, the definition of the crystal changed. The translation symmetry is no more a condition. Only the existence of long range order is required. Of course, the lost of the translation symmetry has severe consequences on the determination of the properties of these so-called aperiodic crystals: where are the atoms?, what about the phase transitions in these crystals?, how to describe the collective vibrations in a crystal with no unit cell and so no Brillouin zone? The goal of this paper is to present the conceptual advances done in these fields up to now. The case of the three families of aperiodic crystals will be treated: incommensurate modulated crystals, aperiodic composites and quasicrystals.

2. Periodic crystals

The periodic crystal is defined by the identical repetition in the three dimensions of space of an elementary unit-cell, characterized by three independent basic vectors a, b and c. The existence of the translation symmetry, defined by the translational vector $\mathbf{T} = n_1\mathbf{a}+n_2\mathbf{b}+n_3\mathbf{c}$, is at the origin of the huge simplification met in solid state physics. Indeed, this brings a problem of 1022 atoms for a crystal of 1 cm^3 to a problem of the few N atoms in the unit cell, N varying from 1 (in Polonium) to some 104 in proteins crystals. The translation symmetry allows the introduction of the following concepts:

- the reciprocal space $\mathbf{G}$, such as $\mathbf{G}.\mathbf{T}= 2\pi$, $\mathbf{G} = \mathbf{G}_{hkl} = h\mathbf{a}^*+k\mathbf{b}^*+l\mathbf{c}^*$ where $\mathbf{a}^*$, $\mathbf{b}^*$ and $\mathbf{c}^*$ are the vectors of the dual space of the real space, called the reciprocal space.
- the structure factor associated to the nodes $\mathbf{G}_{hkl}$ of this reciprocal space, is simply obtained by a sum limited to the N atoms in the positions $\mathbf{r}_j$ in the unit cell: $F(\boldsymbol{G_{hkl}}) = \sum_{j=1}^{N} f_j e^{i\boldsymbol{G}.\boldsymbol{r}_j}$, where f_j is the scattering potential of the atom j.

– the Brillouin zone which is the primitive cell in the reciprocal space. Within this Brillouin zone the collective dynamics of the atoms of the crystal is described in term of 3N phonon branches. Three branches are acoustic branches corresponding in the Brillouin zone center to the 3 Goldstone modes expected in a space of dimension D = 3, and (3N – 3) branches are optic branches. Taking advantage again from the translation symmetry, the description of the electronic bands, the optical and magnetic properties... may also be done within this Brillouin zone.

An important question in solid state physics concerns the phase transformations that crystals may present under external fields (temperature, pressure, electric or magnetic field, light, ...). Describing these phase transitions in matter is determining the cooperative change of a macroscopic system made of a huge number of particles. This problem is of an extreme complexity even in the periodic crystals because the cooperative fluctuations are much longer than the size of the elementary cell. These fluctuations actually diverge in the space and in time at the transition, creating by this way a new macroscopic order. The description of these pretransitional phenomena leading to the symmetry breaking is modelled taking into account:

– the dimension of the physical space.
– the dimensionality of the order parameter (n=1 in the Ising-model, n=2 in the X-Y planer model, n=3 in the Heisenberg model).
– the range of the involved interactions.

These criteria define universality classes. Actually, very few of these models are soluble and, in the 70's, another approach, called group renormalization theory, has allowed the description of the set of variables associated to a phase transition. This method gives access to the values of the critical exponents associated to these different universality classes (Wilson, Fisher, Kadanoff, Nobel Prize 1972). Here, we will not consider such critical approaches, limiting the discussion to the mean field phenomenological approach proposed by Landau and corrected later on by Ginsburg and Levanyuk in order to take into account the pretransitionnal fluctuations [1–5]. The Landau theory does not try to describe the real neighbour-neighbour interaction, but instead introduce a mean effective field applied to a single given particle. In this theory, an order parameter η is defined as the measure of the loss of the symmetry element associated to a symmetry breaking at the transition temperature T_c. Spatial fluctuations of the order parameters correspond to the appearance of a short range order on correlation lengths usually anisotropic ξ_a, ξ_b, ξ_c. To this local order is associated a temporal fluctuation, lifetime τ of these local fluctuations $\eta(\boldsymbol{r},t)$ In the high symmetry phase, $\langle\eta(\boldsymbol{r},t)\rangle_{t=\infty} = \bar{\eta} = 0$ but the fluctuations $\langle\eta^2\rangle$ are non-zero very close to the transition temperature: $\langle\eta(\boldsymbol{r},t)\rangle_\xi \neq 0$ and $\langle\eta(\boldsymbol{r},t)\rangle_\tau \neq 0$. Critical diffuse scattering is measured in the reciprocal space around the critical wave vectors which signs the symmetry breaking. It is described by the Orstein-Zernicke expression, Fourier transform of the pair correlation function:

$$S(\boldsymbol{q},T) \sim \frac{k_B T}{1 + \xi_a^2 q_x^2 + \xi_b^2 q_y^2 + \xi_c^2 q_z^2} \qquad (1)$$

The analysis of experimental profiles of the diffuse scattering $S(\boldsymbol{q},T)$ allows the extraction of the experimental correlation lengths. The lifetime of these correlations is measured at these same locations in the reciprocal space, usually by inelastic neutron scattering techniques.

3. Aperiodic crystals

3.1 Structural aspects

If we limit to structural aspects, there exists three families of aperiodic crystals: the incommensurate modulated phases, the aperiodic composites and the quasicrystals. The loss of the periodicity of a crystalline system has the major consequence that the theoretical models defining the physical properties cannot be based on the Bloch theorem. To recover the lost periodicity, the concept of superpace crystallography has been introduced by de Wolf, Janner and Janssen [6–9]. In a general way, the superspace has a dimension, or rank, n = D+d, where D is the dimension of the physical space (lower or equal to 3) and d the dimension of the internal space of the superspace. The cost of such creation is that the atom is no more defined by the three coordinates $\mathbf{r}_j$ (x_j, y_j, z_j) of the physical space but becomes a function in a space of dimension larger than the physical one. In a space of dimension 4, this function is a curve whereas it is a plane and a volume in spaces of rank respectively 5 and 6 [10, 11]. These different concepts are presented in several books [10, 12, 13].

3.1.1 The incommensurate modulated phases

The modulated incommensurate compounds can be seen as a perturbation of a periodic tridimensional lattice by a modulation whose period is in an incommensurate ratio with the periodicity of the mean lattice. In a general way, the incommensurate phase appears when there is a frustration between antagonistic forces. The first case observed experimentally concerned a magnetic modulation [14]. They correspond to crystals in which a spin wave exists and presents a periodicity which is incommensurate with the period of the lattice. Since that time, a very large number of atomic modulations have been observed [15] and it appears that in fact these phases are rather current in nature. An important research axis in the incommensurate modulated crystals concerns compounds of low dimensionality presenting Peierls instability. In that case, the charge density waves result from the competition between the elastic energy and the electronic energy related to the partial filling of the Fermi band [16]. To illustrate an incommensurate modulated material, let us consider a chain of atoms of period a, modulated in position by a sinusoidal function $A\sin(qx+\phi)$ where $q=\frac{2\pi}{\lambda}$ and $\frac{\lambda}{a}$ is irrational. ϕ is the phase of the modulation. By introducing a perpendicular direction to the physical space, we can create along the new direction a function which is the one of the incommensurate modulation. The atom is no more a point but is defined by this atomic function. The position of the atoms in the physical space is obtained by a three-dimensional cut of a superspace of dimension n (in the Fig. 1, one-dimensional cut of superspace of rank 2). The perpendicular space contains the information on the phase of the modulation and any fluctuation along this direction is called phase mode or "phason". In these aperiodic materials, the term phonon is reserved to vibration modes contained in the physical space. In the extreme simplest case, the atomic function is a sinusoidal function. In real crystals, the function is more complex (described then by the presence of harmonics) and can even become discontinuous when position of the atoms are forbidden for instance for steric reasons. Often, one has considered the discontinuous characters of the atomic functions as specific to quasicrystals, which have indeed an image made of limited segments in the crystallographic superspace. The possible existence of an analycity breaking in incommensurately modulated crystals and in aperiodic composites proves that this property is not limited to quasicrystal [17, 18]. The reciprocal space, dual space of the superspace of dimension n, is also of rang n and is defined by the generalization of the expression $\boldsymbol{G}.\boldsymbol{T}=2\pi\ \boldsymbol{a}_i.\boldsymbol{a}_j^*=2\pi\delta_{ij}$ with $i,j=1,n$.

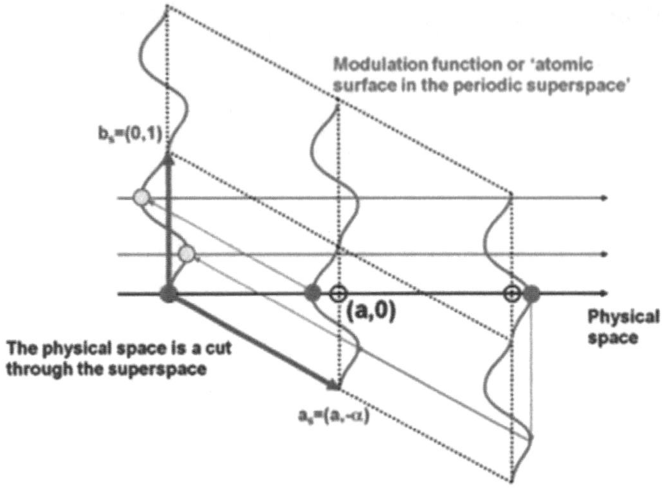

Figure 1. Left: schematic representation of a superspace of dimension 1 + 1 in the real space, the orange curve represent the atomic function along the internal dimension of the superspace. The actual positions of the atoms are given by a horizontal cut; right: in the reciprocal space, the actual Bragg positions are obtained by a projection of the reciprocal nodes on the physical reciprocal space.

In the case of incommensurately modulated crystals, there exists a mean structure which is periodic in three dimensions. This allows the definition of a mean Brillouin zone. In the case of an incommensurate modulation characterized by a single vector of modulation, the dimension of the superspace group is four, characterized by the four reciprocal vectors **a***, **b***, **c*** and $\mathbf{q}_s$. The vectors of the reciprocal space are written:

$$\boldsymbol{G}_{hklm} = h\boldsymbol{a}^* + k\boldsymbol{b}^* + l\boldsymbol{c}^* + m\boldsymbol{q}_s \tag{2}$$

In the reciprocal space, one defines two types of Bragg peaks:

- the Bragg peaks (hkl0) also called main reflections, associated to the mean structure.
- the satellites (hklm) with m≠0 associated to the modulation.

The incommensurate modulated phases results usually from a continuous phase transition. As a consequence, the amplitude A of the modulation is weak at least close to the transition temperature. The calculation of the intensity of the satellite peaks (hklm) with m ≠ 0 can be done in simple cases. Let us consider a sinusoidal modulation as described above, the position the jth atom in the unit cell is:

$$\boldsymbol{U}_j^l = \boldsymbol{R}_l + \boldsymbol{r}_j + \boldsymbol{A}\sin(\boldsymbol{q}.\boldsymbol{R}_l + \varphi_j) \tag{3}$$

For such a crystal the structure factor is:

$$F(\boldsymbol{Q}) = \sum_{j,l} f_j e^{i\,\boldsymbol{Q}.\boldsymbol{U}_j^l} = \sum_{j,l} f_j e^{i\,\boldsymbol{Q}.(\boldsymbol{R}_l + \boldsymbol{r}_j)} e^{i\,\boldsymbol{Q}.\boldsymbol{A}(\sin(\boldsymbol{q}.\mathbf{R}_l + \phi_j)} \tag{4}$$

According to Auger-Jacobi formula:

$$e^{ix\sin(\varphi)} = \sum_{n=-\infty}^{+\infty} J_n(x)\, e^{in\varphi} \tag{5}$$

Table 1. Notation of the superspace groups.

Symbol of the space group of the mean structure	Symbol of the modulation vector, expressed between parenthesis
Example of Na2CO3 in dimension (3 + 1): C2/m ($\alpha 0 \gamma$).	

it comes:

$$F(\boldsymbol{Q}) = \sum_{m=-\infty}^{+\infty} \left(\sum_{j=1}^{N} f_j J_m (\boldsymbol{Q}.\boldsymbol{A}) e^{i\boldsymbol{Q}.\boldsymbol{r}_j} e^{im\phi_j} \sum_{G} \delta(\boldsymbol{Q} + m\boldsymbol{q} - \boldsymbol{G}) \right) \tag{6}$$

The diffracted intensity is then:

$$I(\boldsymbol{G} \pm m\boldsymbol{q}) = \left| \sum_{j} f_j J_m [(\boldsymbol{G} \pm m\boldsymbol{q}).\boldsymbol{A}] e^{i(\boldsymbol{G} \pm m\boldsymbol{q}).\boldsymbol{r}_j} e^{im\phi_j} \right|^2 \tag{7}$$

For small amplitude A of the modulation, $J_m(x) \sim x_m$ and so $(\boldsymbol{G} + m\boldsymbol{q}) \propto [(\boldsymbol{G} + m\boldsymbol{q}).A]^{2m}$ where the term $[(\boldsymbol{G} + m\boldsymbol{q}).A]$ is much lower than 1. So, in theory, there is an infinite number of Bragg peaks distributed uniformly in the reciprocal space. However, since the intensity of the satellite Bragg peaks decrease very fast with their order m, only the satellites with a very low order are observable. The criterion is even more drastic in the case of an occupation modulation, where the intensity of the satellites is non zero only for the order ± 1 [12]. In the general case, however, the modulation is not purely sinusoidal and there are no strict rules concerning the intensity of the different satellites peaks. The diffraction image being constituted of a discrete ensemble of reflections, there exist a finite number of symmetry elements leaving it invariant. The whole set of symmetry operations form the group of superspace of rank D+d. This superspace group is determined by the observation of the reflections conditions discussed here on the D+d indices. So a centering I in a superspace of dimension 4 characterized by the above expression will imply the condition of presence h+k+l+m even [10]. To correctly define the symmetry of an aperiodic crystal, it is essential to identify what are the basic vectors of the system which will define the name of the the superspace group. They will be constituted of the three base vectors **a***, **b*** and **c*** characterizing the physical reciprocal space and of "d" vectors (representing the d dimension of the internal space of the superspace) expressed as a function of these three basic vectors. The notation of the superspace groups is derived from the one of the usual space groups writing close to it the three components of the d modulation vectors.

For instance, for sodium carbonate, the first compound studied by de Wolff et al., the mean structure is monoclinic C2/m. The modulation is characterized by the vector $\boldsymbol{q}_s = \alpha \boldsymbol{a}^* + \gamma \boldsymbol{c}^*$ with α and γ temperature dependant and equal to 0.182 and 0.318, respectively, at 300K. The superspace group is then C2/m ($\alpha 0 \gamma$). The superspace groups have been listed at dimension D=3+1, 3+2 and 3+3. There are 756 superspace groups for modulated structures of dimension D=3+1. They can be found at the following internet address: `http://stokes.byu.edu/iso/ssg.php`.

3.1.2 The aperiodic composites

The aperiodic composites are formed by the intergrowth of two or more sublattices. In these crystals, aperiodicity may appear quite naturally when the lattice periodicities of the subsystems cannot adjust to each other at least in one direction. Well-known examples are the inorganic misfit layers (Fig. 2), two dimensional "sandwich" compounds where in different layers the parameters do not match [12]. The aperiodic composite can be seen as the imbrication of two incommensurately modulated sub-systems. Each subsystem plays

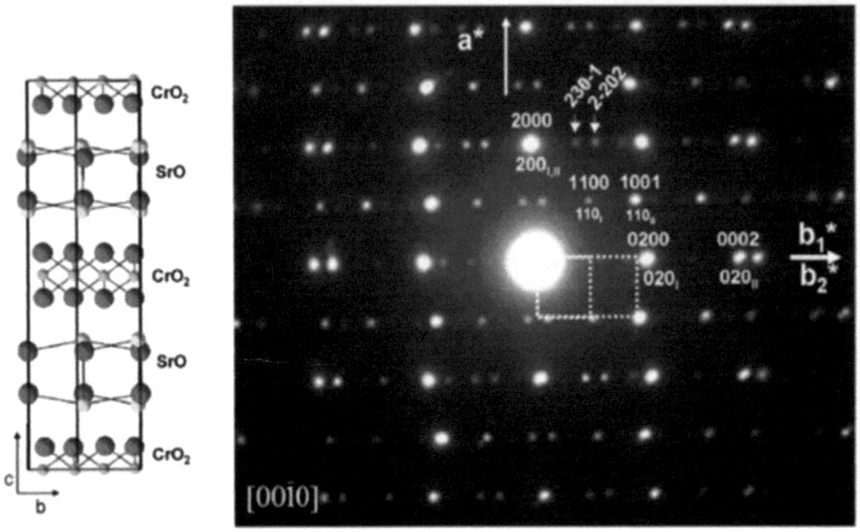

Figure 2. A two dimensional composite structure $[Sr_2O_2][CrO_2]$ 1.85 with its electronic diffraction image indexed with 4 indices [19].

the role of an external potential for the other(s) sub-system(s). The major difference with an incommensurate modulated crystal is that aperiodic composites are usually aperiodic by construction: there is no high symmetry periodic phase, no mean periodic structure and no mean Brillouin zone.

In the case of a single incommensurate direction, the crystal must be described within a superspace of rank 3+1 [20]. The lattice possesses four characteristic periodicities **a**, **b**, $\mathbf{c}_1$ and $\mathbf{c}_2$, where **a** and **b** are the two vectors of the commensurate plane (perpendicular to the aperiodic direction, $\mathbf{c}_1$ and $\mathbf{c}_2$ the basis vectors of the two sublattices 1 and 2 along the incommensurate direction. Some examples of prototype materials are listed below and illustrated in the Fig. 3:

- the mercury chains in an AsF_6 lattice, $Hg_{3-\delta}AsF_6$ with δ of the order of 0.20 [21].
- "self hosting" composites where the same atoms in host and guest, sublattices [22].
- the tetrathiafulvalene salts TTF-Br_x and TTF-I_x with x of the order of 0.79 and 0.71 respectively [23, 24].
- host/guest inclusion compounds of urea/alkane.

In the superspace, the crystal pattern is the imbrication of two atomic surfaces associated to the atoms of the two lattices. In order to describe the whole set of reciprocal vectors of the reciprocal space, one needs four reciprocal basic vectors $\mathbf{a}^*$, $\mathbf{b}^*$, $\mathbf{c}_1^*$ and $\mathbf{c}_2^*$. Contrary to the case of incommensurate modulated crystals, there is no more the notion of main reflection. In the present case, there are four types of Bragg peaks (Fig. 4):

- (hk00): common Bragg peaks defining the commensurate reciprocal plan ($\mathbf{a}^*$, $\mathbf{b}^*$).
- (hkl0): Bragg peaks reflecting the periodicity the first sublattice 1.
- (hk0m): Bragg peaks reflecting the periodicity the second sublattice 2.
- (hklm) with l and $m \neq 0$: intermodulation satellite Bragg peaks appearing at combinatory positions of $\mathbf{c}_1^*$ and $\mathbf{c}_2^*$ and reflecting the intermodulation between the two sublattices.

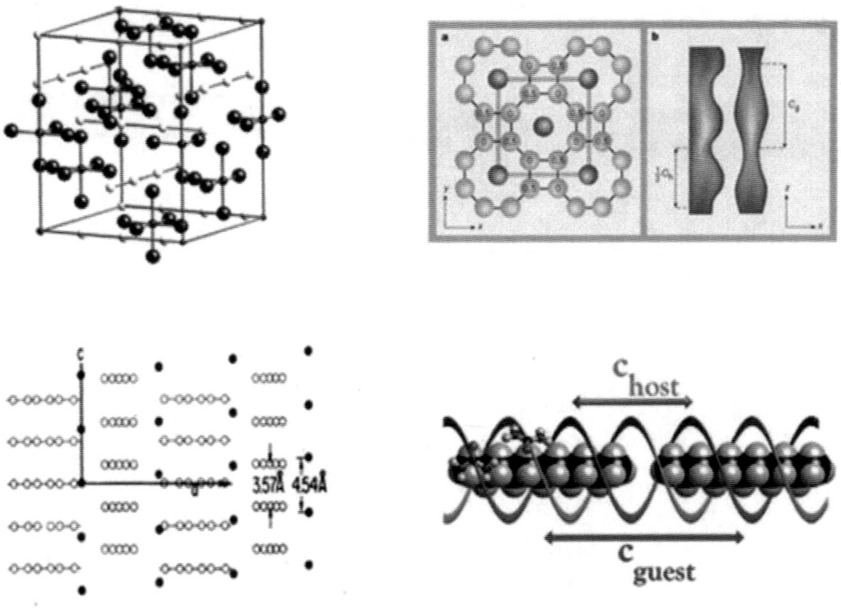

Figure 3. Examples of aperiodic composites $Hg_{3-\delta}AsF_6$, Barium self-hosted in its phase "IV", tetrathiafulvalene salt TTF-Br_x and urea/n-alkane inclusion compound.

In fact, both sublattices contribute to the intensity of the whole set of Bragg peaks. This makes the structure resolution much more complicated than in the case of incommensurate modulated crystals.

For the determination of the superspace group of an aperiodic composite, the first step is to determine the pseudo-space groups and superspace groups associated to the different sub-systems considered separately [25]. The superspace group associated to the whole composite will be the largest common subgroup between the superspace groups of the different sub-systems. The complete list of superspace groups for uniaxial aperiodic composites of dimension $3 + 1$ has been given by van Smaalen and Harris [26].

Since aperiodic composites are aperiodic by constructions, they are very much adapted to study symmetry breakings within crystallographic superspaces. The prototype family of alkane/urea compounds has revealed very original sequences of phases in these crystals: a phase transition doubling only the internal space of the superspace by increasing the dimension of the crystallographic superspace [27, 28], the creation of monoclinic intermodulated aperiodic composites [29]. The use of combined coherent neutron scattering and high resolution synchrotron studies permits the analysis of very original pretransitional phenomena within the crystallographic superspace [30].

3.1.3 Quasicrystals

During a long period, crystallography considered crystals as a covering of the space resulting from the arrangement of one type of polyhedra, copied then to infinity by translation. This rule constrains the unit cell to have rotational symmetry of the order 2, 3, 4 and 6. However, more complex complex ways of covering the space have been imagined. A nice illustration is given by the Fibonacci chain where one considers an atom A and an atom B. A trivial periodic tiling is obtained by simple juxtaposition of the object AB: ABABAB. A more complex process,

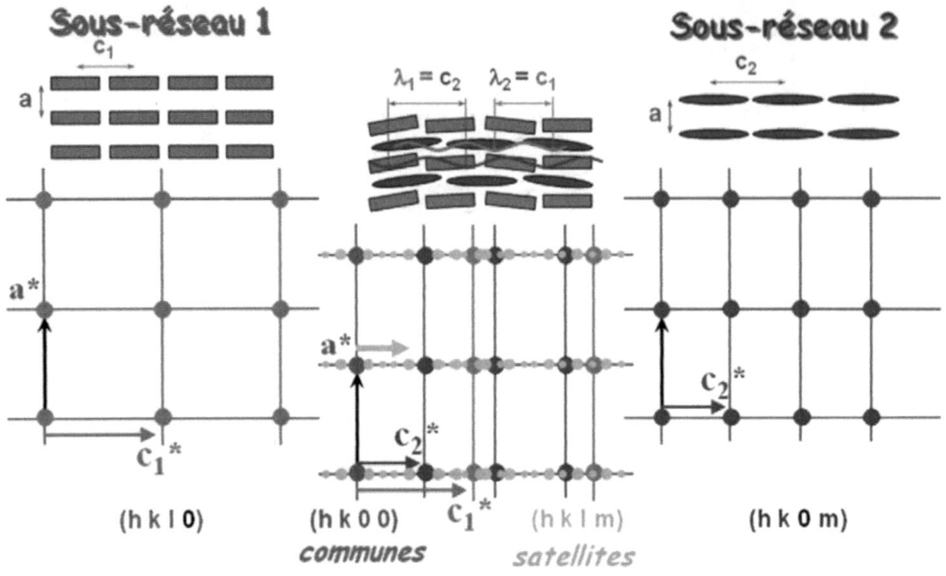

Figure 4. A composite crystal results from at least two imbricated and intermodulated sublattices 1 and 2. The diffraction image is not the simple superposition of the two independent diffraction images (right and left) but much more Bragg peaks, satellites in green, are generated by the intermodulation.

called inflation rule, where each arrangement S(n) of the step n is the juxtaposition of the arrangements S(n-1) and S(n-2), yields a long range order but without translation symmetry (Fig. 5):

A, AB, ABA, ABAAB, ABAABABA, ABAABABAABAAB, etc..

Fibonacci chain appears as a cut of a square lattice plus projection. It is easily shown that the ratio of the number of the atoms A and B is equal for an infinite chain to the golden number $\tau = 2\cos\left(\frac{2\pi}{10}\right)$. The two dimensional equivalent is given by a tiling with two lozenges of base $\frac{2\pi}{10}$ and $\frac{2\pi}{15}$. This tiling created by R. Penrose in 1978 is also non-periodic. It can be shown that it is the intersection of a four dimensional periodic superspace. The three dimensional representation is closed to the one encountered in some real quasicrystals. These quasicrystals may present five-fold (pentagonal), 8 fold (octagonal), 10 fold (decagonal) or 12 fold (dodecagonal) symmetries (Fig. 6). More complexes tiling may also exist [13]. The experimental discovery of quasicrystals in 1984 was presented in the Physical Review letters with the title: "Metallic Phase with long range orientational order and no translational symmetry" by D. Schechtman, I. Blech, D. Gratias and J. W. Cahn. It was awarded by the Nobel Prize of Chemistry in 2011. We know today hundred quasicrystalline structures, mainly ternary metallic alloys often with aluminium. These quasicrystals are described within superspaces of rank up the 6. Solving the structure of a quasicrystal is determining the shape of the discontinuous atomic hypervolumes [31, 32]. The problem is of course very complex. The study of approximant commensurate phases allows sometime to tackle the problem in a simpler way [33]. Another efficient approach is to modelise the structure of a quasicrystal by a limited number of elementary blocks or "clusters" [34, 35].

The existence of diffuse scattering in the quasicrystal diffraction image is quite usual. Of course, there is the same type of static (substitution, occupation and displacement) and dynamic (phonons and atomic diffusion) disorder than in usual periodic crystals. Other

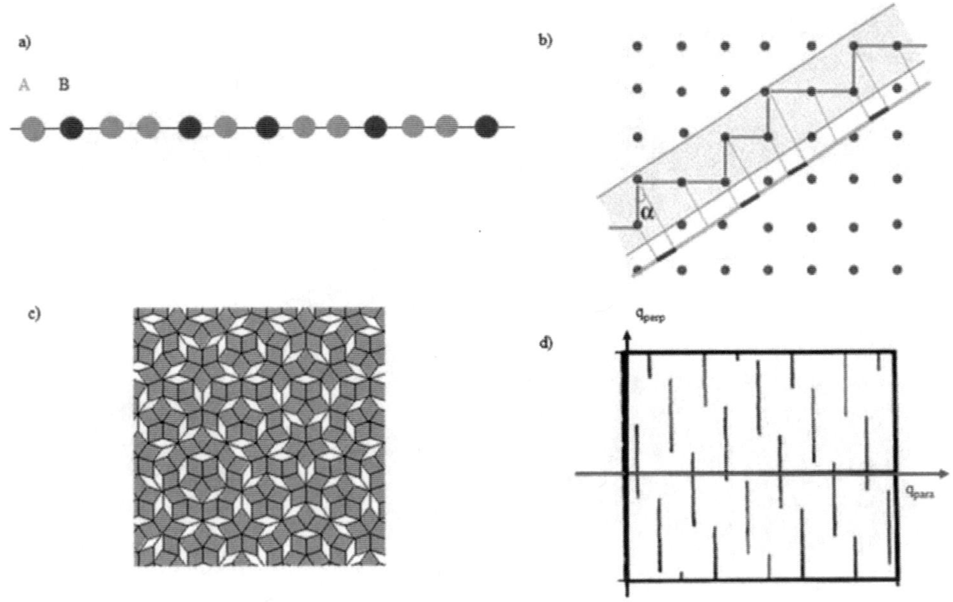

Figure 5. a) Fibonacci chain of atoms A and B;. b) A 1D aperiodic Fibonacci chain is obtained by a cut of a periodic two dimensional square lattice; c) A Penrose tiling, the two dimensional analog of the Fibonacci chain; d) the atomic function corresponding to a quasicystal is given by finite segments along the internal space of the crystallographic superspace.

kinds of disorder are specific to quasicrystals and are related to the difficulty in defining a simple mean structure. Let us note that until now, there is a discussion on the origin of the stabilization of the quasicrystals either by the energy or by the entropy. In that last case, the quasicrystals would transform in a periodic approximant on approaching zero Kelvin, whereas in the first case the quasicrystalline state would be the fundamental state. The very slow kinetic of the metallic structures at low temperature does not allow to answer this question. This complexity appears in all studies of phase transitions in these materials, and consequently few examples illustrate the broken symmetry in quasicrystals.

3.2 Dynamics in aperiodic crystals

3.2.1 Incommensurate modulated crystals

The incommensurate modulated crystals phases are the simplest aperiodic crystals and they were extensively studied in the 80's [15]. The simplicity in these materials comes from the fact that they have a high symmetry periodic phase. The Brillouin zone of these periodic crystals will play the role of the mean Brillouin zone in the incommensurate phase. In that phase, a huge majority of the phonon branches remain unchanged compared to the periodic crystal. The Fig. 7 summarizes the structural characteristics of an incommensurate crystal which possesses a high symmetry periodic phase and a lock-in low temperature commensurate phase.

The basis from the intermediate incommensurate phase of rank four, characterized by an unique critical modulation vector $\mathbf{q}_s$, is described by the vectors $\mathbf{a}^*$, $\mathbf{b}^*$, $\mathbf{c}^*$ and $\mathbf{q}_s$. The static modulation appearing in the incommensurate phase, couples the phonon branches degenerated in $\pm\mathbf{q}_s$, giving rise to two new modes a symmetric one, the amplitudon, and an asymmetric one, the phason [37]. Referring to displacive phase transition governed

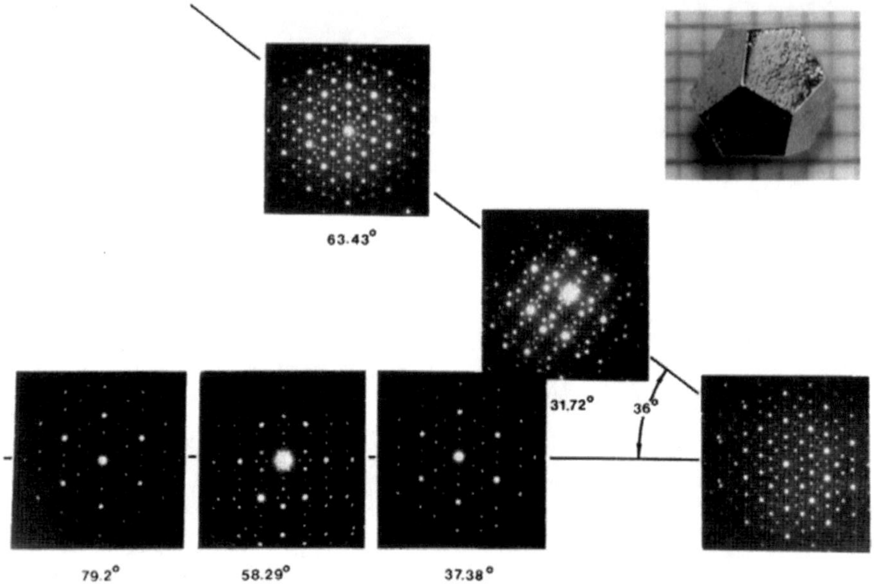

Figure 6. Electronic diffraction image of an icosahedral quasicrystal [36]. In inset, an icosahedral quasicrystal of Zn-Mg-Ho.

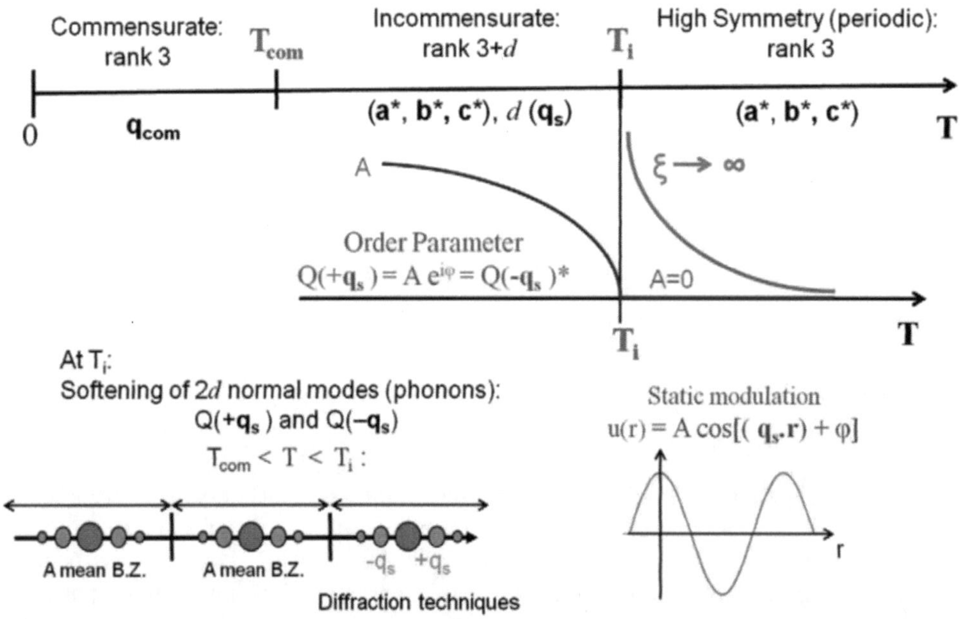

Figure 7. Schematic representation of the sequences of phase in a modulated incommensurate crystal which possess a periodic high symmetry phase. At the transition temperature T_i, the correlation lengths ξ diverge and below T_i one defines an order parameter as a complex quantity $Q(\mathbf{q}_s) = Ae^{i\varphi} = Q(-\mathbf{q}_s)^*$. The incommensurate phase is characterized by the static modulation $u(r) = A\cos[(\mathbf{q}_s.r)+\varphi]$ where A is the amplitude and φ the phase of the modulation. The existence of the static modulation induces the existence of satellite Bragg peaks at positions $\pm nq_s$ (in green) around the high symmetry Bragg peaks (in red). Often, at lower temperature, there is a lock-in phase transition towards a commensurate phase at T_{com}.

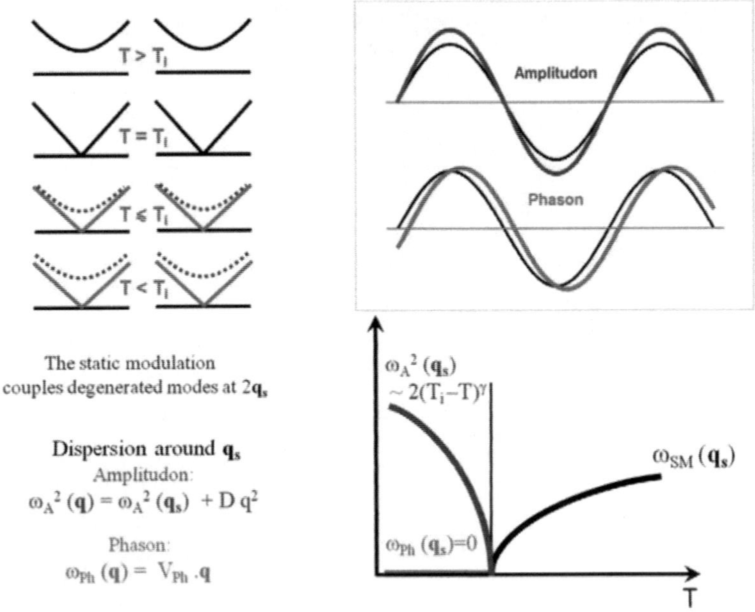

Figure 8. Collective dynamics around a displacive phase transition to an incommensurate modulated phase. Below T_i, the static modulation couples degenerated modes separated by $2\mathbf{q}_s$ creating two excitations the amplitudon (usual soft mode) in blue and the phason (zero energy mode at $\mathbf{q}_s$) in red.

by a soft phonon branch in the high symmetry phase (Fig. 8), the amplitudon within the incommensurate phase is the usual low symmetry soft mode. In the same way, it presents the following dependencies in $\mathbf{q}_s$ and around $\mathbf{q}_s$: $\omega_A^2(\boldsymbol{q}_s) \propto 2\,(T_i - T)^{\gamma}$ and $\omega_A^2\,(q) = \omega_A^2(\mathbf{q}_s) + Dq^2$. The associated phason mode presents the linear dispersion, going from zero energy at $\mathbf{q}_s$ in the whole incommensurate phase $\omega_{Ph}^2(\mathbf{q}_s) = 0$ and $\omega_{Ph}(\boldsymbol{q}) = V_{Ph}.\boldsymbol{q}$. The phason mode corresponds to a fluctuation along the internal space of the superspace. It is like a supplementary Goldstone-like mode with a liner dispersion as an acoustic phonon (in the case of a superspace of rank D= 3+d, d such branches are predicted). It should be mentioned that even if these modes present a linear dispersion in zone center within the whole incommensurate phase, they have a finite damping at $\mathbf{q}_s$ contrary to real Goldstone modes (acoustic mode at Bragg position). The reason is that these phason modes correspond to a relative displacement of the atoms (which means dissipation) which is not the case for acoustic phonons. Experimentally, this last property explains why phason modes are so efficient to the spin-lattice relaxation rate in NMR spectroscopy and indeed many incommensurate phases where discovered by this technique. However, coherent neutron scattering was the unique way to measure the phason spectral function getting by the way frequency dispersion and damping of this excitation [15, 38–41].

3.2.2 Dynamics in aperiodic composites

The collective dynamics in aperiodic composite is very complex because of two main reasons. The first one is the absence of a high symmetry periodic phase, which makes impossible the calculation for a finite number of atoms as in incommensurate modulated crystals. The second reason is that it is impossible to define even a mean Brillouin zone, and so the real Brillouin zone is effectively of zero volume. Such a result is particularly well illustrated by

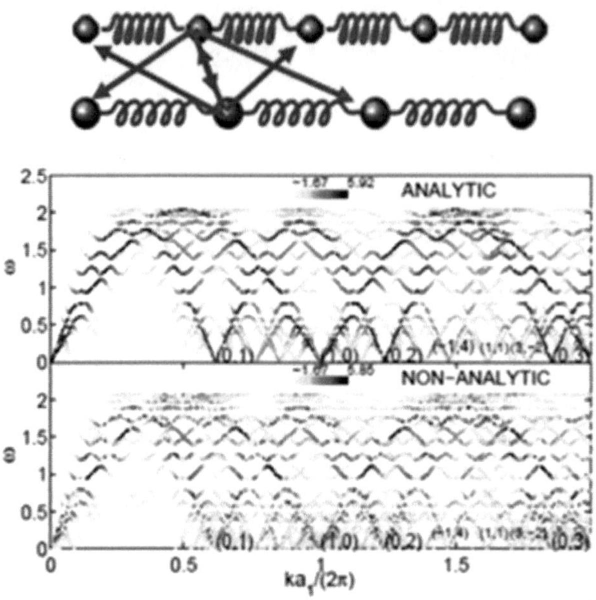

Figure 9. Collective dynamics in an aperiodic composite crystal modelled by coupled deformable chains (up). The phonon branches are shown in the case an analytic and of a non-analytic atomic function. The colors show the dynamical structure factor on a logarithmic scale [42].

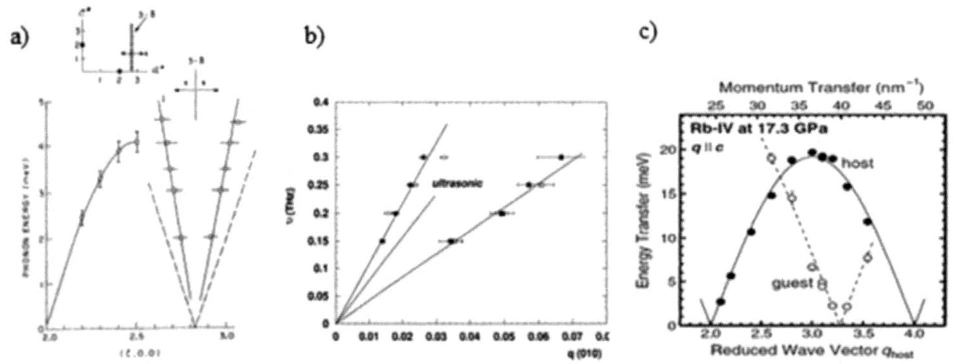

Figure 10. Longitudinal acoustic phonon branches measured along the aperiodic direction, a) in mercury chain $Hg_{3-\delta}AsF_6$, b) in $Bi_2Sr_2CuO_{6+x}$ and c) in the "self-hosting" crystal of rubidium in its phase "IV".

the calculation of deformable coupled aperiodic chains by T. Janssen et al. [42], as shown in the Fig. 9.

The analytical resolution of the equation of motion leads the dispersion branches shown in the figure. Two properties come out: the emergence of phonon branches out of the four type of Bragg peaks defined previously for a one dimensional aperiodic composite (hk00), (hkl0), (hk0m) and (hklm). Aperiodicity can be seen as resulting from an infinite folding of the dispersion branches with multiple energy gaps opening in these branches. The curves obtained experimentally appear much more simples than the ones predicted theoretically (Fig. 10). The reason is that the inelastic structure factors are in most case too weak and only allows the observation of the phonon branches emerging from structure common, and each

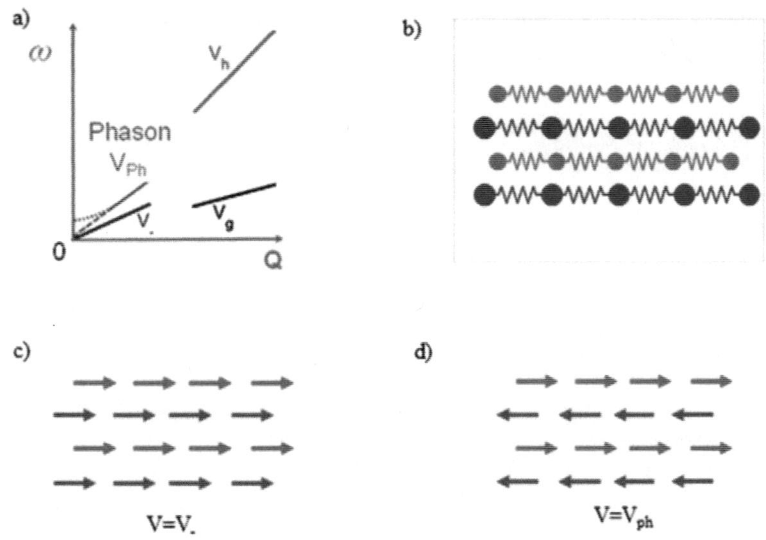

Figure 11. (a) Dispersion branches of two sublattices (V_h and V_g) coupled close to Q=0 to give acoustic phonon branch of slope V_ and a phason branch of slope V_{ph}, without gap in energy if the atomic function is analytical (b), (c) et (d) Schematic representation of the components of the motions in phase for the acoustic phonon (a unique speed of sound in the crystal), in antiphase for the phason ("sliding mode").

sublattice Bragg peaks. In some way, the obtained image experimentally corresponds solely to the signature of the collective dynamics of the two quasi-independent sublattices ignoring the influence of their intermodulation. So, each Bragg peaks yields the elastics constant of its on sublattice, force constant defining the slope of the longitudinal acoustic phonon branch in the incommensurate direction. Two rather different slopes were observed in Bi_2Sr_2CuO [43]. In the "self-hosting" composite, where the same atom is implicated both in the host and in the guest matrix, it is quite naturally the same slope which is measured around (hkl0) and (hk0m) Bragg peaks [44]. In the high symmetry phase of $Hg_{3-\delta}AsF_6$, the mercury presents a one dimensional liquid-like character to which are associated scattering plane perpendicular to the incommensurate direction (in this case, actually two orthogonal directions). The coherent neutron scattering measurements has permitted the observation of two branches with totally different slopes associated respectively to the host AsF_6 matrix and the guest quasi-liquid mercury [45].

Axe and Bak have shown that at the origin of the reciprocal space of an aperiodic composite the two branches of the host and guest couple together to generate two new excitations [42, 46, 47]. One branch is the usual acoustic branch corresponding to an in phase displacement of the atoms of both subsystems. The slope of this branch defines the unique speed of sound in these aperiodic crystals. The other branch is associated to fluctuations along the internal space of the crystallographic superspace and is a phase fluctuation mode (Fig. 11). This antisymmetric mode keeps the center of mass of the two sublattices when vibrating in antiphase. In aperiodic composites, this phason mode is usually called "sliding mode". No direct observation of such a sliding mode has been done yet.

A recent study [48] reported the coupling of the longitudinal acoustic mode of the host subsystem with such a sliding mode (Fig. 12). Measurement was performed by neutron scattering. It was done on an organic host-guest compound *n*-nonadecane/urea. Phonon branches were measured along the incommensurate direction close to a strong host (urea)

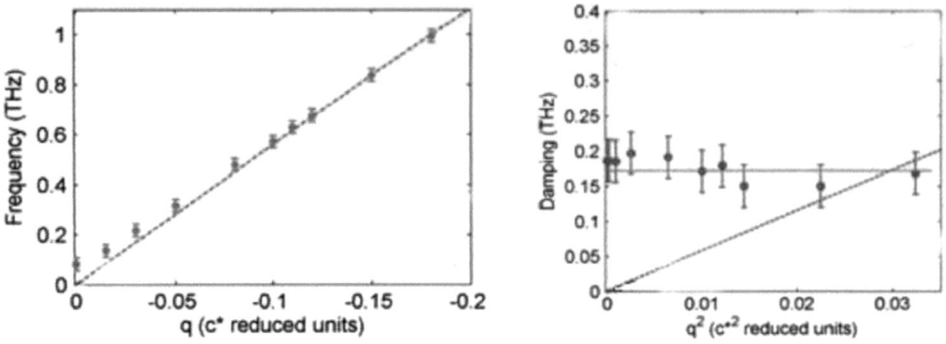

Figure 12. Neutron study of the coupling of the longitudinal acoustic mode of the host subsystem with a sliding mode in *n*-nonadecane/urea. It reveals an energy gap on the L.A. phonon branch together with a non-vanishing damping of this mode on approaching the Bragg position [48].

Bragg peaks. It reveals two new features: an energy gap on the L.A. phonon branch together with a non-vanishing damping of this mode on approaching the Bragg position. This measurement is an indirect evidence of the influence of the phason mode on usual acoustic phonons in an aperiodic crystal. Much more studies have to be done in aperiodic composites before getting a full understanding of the collective dynamics in such crystals without Brillouin zone.

3.2.3 Dynamics in quasicrystals

The determination of the collective dynamics in the quasicrystals is certainly one of the most complex problems in crystals. So, the calculation of the phonon branches is more easily done by considering commensurate approximants. The Fibonacci chain appears as a simple one dimensional representation of a quasicrystal. Its diffraction image is made of Bragg peaks and the phonon branches appear like emerging from the most intense Bragg peaks, which somehow play the role of zone center Γ [49]. The simplest periodic approximation is the simple repetition here of long L and short S segment: LSLSLS. The dispersion branches corresponding to this arrangement is presented in the Fig. 13. Such an image can be generalized to quasicrystals where Bragg peaks of 5 and 10 fold symmetry could play the role of zone center where acoustic phonon branches will emerge. Reciprocal image of a quasicrystal requires the use of 4 to 6 indices depending on the dimension of its crystallographic superspace. Again, physical space and internal space (often called perpendicular space) are defined. Around the nodes of the physical reciprocal space is measured the diffuse scattering called "Thermal Diffuse Scattering" (TDS) associated to phonons. Around the nodes of the perpendicular space is measured the diffuse scattering called "Phason Diffuse Scattering" (PDS) associated to phason modes. The nature of quasicrystals, aperiodic materials described in crystallographic superspaces by discontinuous atomic surfaces, generates other types of dynamics and disorder in addition to the hydrodynamics modes relative to TDS and PDS. So, diffusive modes associated to atomic jumps ("dynamic phason flip") where evidenced by local probes such as N.M.R. and incoherent neutron scattering [50, 51]. An extensive study of phonon and phason modes in quasicrystals where realized by M. de Boissieu (Fig. 14) and other teams. Such experimental studies require large prototypical quasicrystals. These results are discussed in several books and articles.

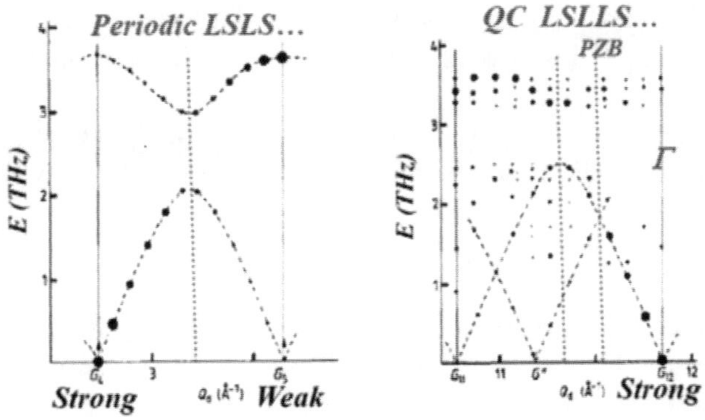

Figure 13. Dispersion branches for a periodic crystal (left) and for a quasicrystal described by a Fibonacci chain (right). In that last case, the strong Bragg peaks play the role of a center of zone Γ of a pseudo Brillouin Zone (PZB) [49].

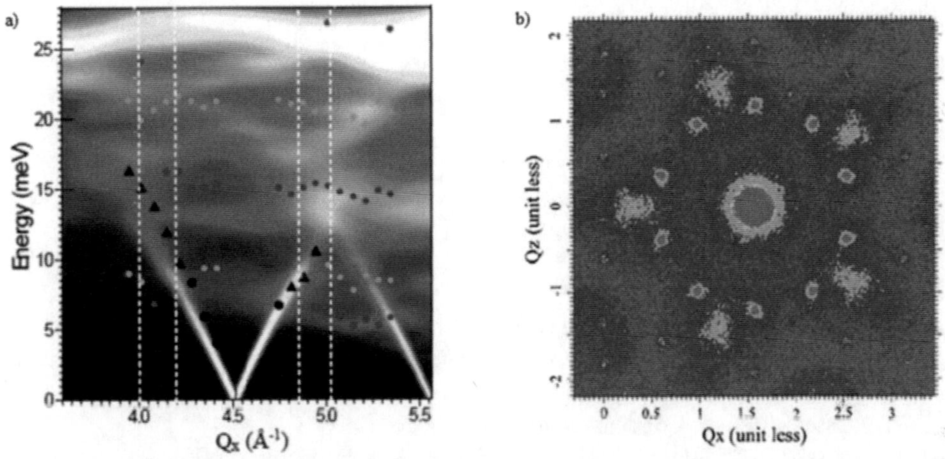

Figure 14. a) Longitudinal modes measured by X-ray inelastic scattering in the quasicrystal ZnMgSc [52] b) Diffuse scattering in the plane perpendicular to the 5 fold symmetry of the i-AlPdMn [53].

4. Conclusion

Neutron scattering has played a major role in the understanding of the order and the dynamics in solid state matter since the middle of the twentieth century. Since the discovery of aperiodic crystal neutron diffraction coupled to X-ray and synchrotron diffraction has allowed the generalization of the standard crystallography towards superspace crystallography. It appears then that a periodic crystal is a particular case of a more general definition of the crystal. It is obtained when the number independent vectors needed to described the periodic cell is equal to the dimension of the physical space (we then say that the rank of the space is equal to the dimension of the space). In aperiodic crystals the rank is larger than the physical space. The crystallographic superspace then decomposes in two spaces the physical space and the internal space (sometime called perpendicular space). Within this superspace, translation symmetry is recovered, defining a supercell and complex atomic functions within this supercell. The structural solution results from a physical cut of this high dimensional space. Many more space groups can be generated by increasing the rank of the superspace.

This offers to the aperiodic crystal an increase of the possible structural solution. Neutron diffraction appears there as an essential technique to study the phase transition within crystallographic superspaces. These phase transitions are associated to critical phenomena and critical dynamics. Coherent neutron scattering is the most adapted technique to measure the collective excitation in these materials which do not possess any Brillouin zone.

References

[1] L. Landau, E. Lifchitz, *Physique Statistique* (Editions MIR, Moscou, 1964)
[2] R. Cowley, Adv. Phys. **29**(1), 1–110 (1980)
[3] A. Bruce, Adv. Phys. **29**, 111 (1980)
[4] A. Bruce, R. Cowley, Adv. Phys. **29**, 219–321 (1980)
[5] A. Levanyuk, Sov. Phys. Solid State **5**, 1294–1298 (1964)
[6] P.M. de Wolff, Acta Cryst. A **30**, 777 (1974)
[7] P.M. de Wolf, Acta Cryst. A **33**, 493–497 (1977)
[8] P.M. de Wolff, T. Janssen, A. Janner, Acta Cryst. A **37**, 625–636 (1981)
[9] T. Janssen, Acta Cryst. A **68**, 667–674 (2012)
[10] T. Janssen, G. Chapuis, M. de Boissieu, *Aperiodic Crystals, From modulated phases to quasicrystals* (Oxford science publication, International Union of crystallography, 2007)
[11] S. van Smaalen, Z. für Kristallogr. **219**, 681–691 (2004)
[12] S. van Smaalen, *Incommensurate Crystallography* (Oxford Science Publication, International Union of Crystallography, 2009)
[13] D. Gratias, R. Mosseri, J. Prost, J. Toner, M. Duneau, *Du cristal à l'amorphe* (Éditions de Physique, Les Ulis, 1988)
[14] A. Herpin, P. Mériel et J. Villain, Compt. Rend. Acad. Sciences **249**, 1334 (1959)
[15] R. Blinc, A. Levanyuk, *Incommensurate phases in dielectrics* (North Holland, 1986)
[16] P. Monceau, Adv. Phys. **61**, 325–581 (2012)
[17] T. Janssen, J. Phys.: Condens. Matter **14**, 12411 (2002)
[18] T. Janssen, O. Radulescu, A. Rubtsov, Eur. Phys. J. B **29**, 136 (2002)
[19] E. Castillo-Martinez, A. Schönleber, S. van Smaalen, A.M. Arevalo-Lopez, M.A. Alario-Franco, J. Solid State Chem. **181**, 1840–1847 (2008)
[20] P. Coppens, Acta Cryst. B **51**, 402–410 (1995)
[21] J.P. Pouget, G. Shirane, J.M. Hastings, A.J. Heeger, N.D. Miro, A.G. MacDiarmid, Phys. Rev. B **18**, 3645–3656 (1978)
[22] R.J. Nelmes, D.R. Allan, M.I. McMahon, S.A. Belmonte, Phys. Rev. Lett, **83**, 4081–4084 (1999)
[23] J.B. Torrance, B.A. Scott, B. Welber, F.B. Kaufman, P.E. Seiden, Phys. Rev. B **19**, 730 (1979)
[24] C.K. Johnson, C.R. Watson, J. Chem. Phys. **64**, 2271 (1976)
[25] S. van Smaalen, Phys. Rev. B **43**, 11330 (1991)
[26] S. van Smaalen, K.D.M. Harris, Superspace group description of the symmetries of incommensurate urea inclusion compounds, *Proceedings: mathematical, physical and engineering sciences*, p. 677–700 (1996)
[27] B. Toudic, P. Garcia, C. Odin P. Rabiller, C. Ecolivet, E. Collet, P. Bourges, G.J. McIntyre, M.D. Hollingsworth, T. Breczewski, Science **319**, 69 (2008)
[28] B. Toudic, P. Rabiller, L. Bourgeois, M. Huard, C. Ecolivet, G.J. McIntyre, P. Bourges, T. Breczewski, T. Janssen, Temperature-pressure phase diagram of an aperiodic host guest compound, Europhys. Lett. **93**, 16003 (2011)

[29] C. Mariette, M. Huard, P. Rabiller, S.M. Nichols, C. Ecolivet, T. Janssen, K.E. Alquist, M.D. Hollingsworth, B. Toudic, J. Chem. Phys. **136**, 104507 (2012)
[30] C. Mariette, L. Guérin, P. Rabiller, C. Ecolivet, P. García-Orduña, P. Bourges, A. Bosak, D. de Sanctis, M.D. Hollingsworth, T. Janssen, B. Toudic, Phys. Rev. B **87**, 104101 (2013)
[31] C. Janot, J-M. Dubois, M. de Boissieu, Am. J. Phys. **57**, 972 (1989)
[32] D. Gratias, M. Quinquandon, Philosoph. Mag. **8**, 1887–1903 (2008)
[33] A. Yamamoto, H. Takakura, Phys. Rev. B **68**, 132201 (2003)
[34] D. Gratias, F. Puyraimond, M. Quinquandon, A. Katz, Phys. Rev. B **63**, 024202 (2000)
[35] M. Quinquandon, D. Gratias, Phys. Rev. B **74**, 214205 (2006)
[36] D. Schechtman, I. Blech, D. Gratias, J. W. Cahn, Phys. Rev. Lett. **53**, 1951–1953 (1984)
[37] J.D. Axe, Incommensurate Structural Phase Transformations, *Proc. of Gatlinburg Conf. on Neutron Scatt.*, edited by Moon R., p. 353 (1976)
[38] H. Cailleau, Solid State Comm. **33**, 407–411 (1980)
[39] L. Bernard, R. Currat, P. Delamoye, C.M.E. Zeyen, S. Hubert, R. de Kouchkovksy, J. Phys. C: Solid State Phys. **16**, 433 (1983)
[40] P. Launois, F. Moussa, M.H. Lemée-Cailleau, H. Cailleau, Phys. Rev. B **40**, 5042–5055 (1989)
[41] J. Ollivier, J. Etrillard, B. Toudic, C. Ecolivet, P. Bourges, A.P. Levanyuk, Phys. Rev. Lett. **81**, 3667 (1998)
[42] O. Radulescu, T. Janssen, J. Etrillard, Eur. Phys. J. B **29**, 385–398 (2002)
[43] J. Etrillard, L. Bourgeois, P. Bourges, B. Liang, C.T. Lin, B. Keimer, Europhys. Lett. **66**, 246 (2004)
[44] I. Loa, L.F. Lundegaard, M.I. Mc Mahon, S.R. Evans, A. Bossak, M. Krisch, Phys. Rev. Lett. **99**, 035501 (2007)
[45] J.M. Hastings, J.P. Pouget, G. Shirane, A.J. Heeger, N.D. Miro, A.G. MacDiarmid, Phys. Rev. Lett. **39**, 1484–1487 (1977)
[46] J.D. Axe, P. Bak, Phys. Rev. B **26**, 4963 (1982)
[47] R. Currat, E. Kats, I. Luk'yanchuk, Phys. J. B **26**, 339 (2002)
[48] B. Toudic, R. Lefort, C. Ecolivet, L. Guérin, R. Currat, P. Bourges, T. Breczewski, Phys. Rev. Lett. **107**, 205502 (2011)
[49] G. Poussigue, C. Benoit, M. de Boissieu, R. Currat, J. Phys. Condens. Matter **6**, 659 (1994)
[50] S. Lyonnard, *Etude des sauts atomiques dans les quasicristaux*, Thèse de doctorat de l'Université de Paris 11 (1997)
[51] G. Coddens, W. Steurer, Phys. Rev. B **60**, 270 (1999)
[52] M. de Boissieu, S. Francoual, M. Mihalkovic, K. Shibata, A.Q.R. Baron, Y. Sidis, T. Ishimasa, D. Wu, T. Lograsso, L.-P. Regnault, F. Gähler, S. Tsutsui, B. Hennion, P. Bastie, T.J. Sato, H. Takakura, R. Currat, A.-P. Tsai, Nature Mat. **6**, 977 (2007)
[53] M. de Boissieu, S. Francoual, Y. Kaneko, T. Ishimase, Phys. Rev. Lett. **95**, 105503 (2005)

EPJ Web of Conferences **155**, 00005 (2017) DOI: 10.1051/epjconf/201715500005
JDN 22

Crystallographic textures

Vincent Klosek[a]

CEA, IRAMIS, Laboratoire Léon Brillouin, 91191 Gif-sur-Yvette Cedex, France

Abstract. In material science, crystallographic texture is an important microstructural parameter which directly determines the anisotropy degree of most physical properties of a polycrystalline material at the macro scale. Its characterization is thus of fundamental and applied importance, and should ideally be performed prior to any physical property measurement or modeling. Neutron diffraction is a tool of choice for characterizing crystallographic textures: its main advantages over other existing techniques, and especially over the X-ray diffraction techniques, are due to the low neutron absorption by most elements. The obtained information is representative of a large number of grains, leading to a better accuracy of the statistical description of texture.

1. Introduction

A macroscopic physical property is a relation between a macroscopic stimulus and a macroscopic response [1]. In their general form, such relations are tensorial to account for possible anisotropy, and tensorial field variables have to be considered. For instance, electrical conductivity describes the relation between the current density vector **j** and the electric (vector) field **E** by application of a second order tensor, the conductivity tensor $\boldsymbol{\sigma}$. Or elasticity theory establishes the well-known generalized Hooke law, which is a linear relation between the stress $\boldsymbol{\sigma}$[1] and strain ε tensors, which are second order symmetric tensors, introducing the 4th order elastic stiffness tensor **C**.

A single crystal being intrinsically an anisotropic (although periodic) arrangement of atoms, most single crystal physical properties are anisotropic, i.e., they are dependent on the crystallographic direction: isotropy is the exception. For instance, the variations of the Young modulus of a Cu single crystal are plotted on Fig. 1 as a function of the crystallographic direction (along which the uniaxial elastic stress is applied and the elastic strain is measured): Cu is stiffer along <111> directions, the Young modulus being almost three times larger than along <100>. Of course, the anisotropy of a single crystal physical property depends on the crystal symmetry (Sect. 2), but also on the tensorial nature of the stimulus-response relation: for instance, any property which is described by a 2nd order tensor (e.g., the conductivity, the thermal dilatation, etc.) is nevertheless isotropic if the crystal symmetry is cubic. Crystal symmetries and their influence on single crystal physical properties are the subject of several books: for instance, the preliminary study of "classic" Ref. [2] is highly recommended.

A polycrystal is an aggregate of numerous single crystals more or less randomly orientated. Hence, its macroscopic physical properties should be directly influenced by the

[a] e-mail: vincent.klosek@cea.fr

[1] It is unfortunate that standard notation uses the same $\boldsymbol{\sigma}$ symbol for both conductivity and stress tensors.

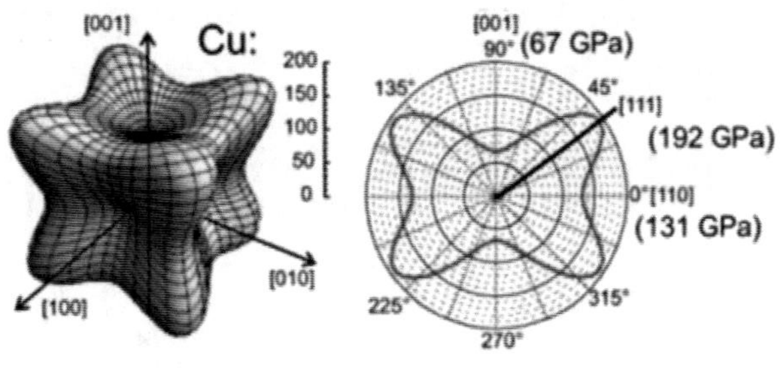

Figure 1. Variation of the Young modulus in a Cu crystal as a function of crystal orientation [3].

local and statistical distributions of grains orientations, as well as by the nature of the interactions between adjacent grains.

Crystallographic texture describes the preferential orientations of the crystallites within a polycrystalline sample, or, in other words, the non uniform distribution of crystallographic orientations. It is thus an important microstructural parameter whose representation can be either statistical or local, either qualitative of quantitative. However, in crystallography, defining an orientation is not trivial and depends on the context: are we dealing with the lattice orientation, the crystal orientation (i.e., the orientation of the crystalline motif), the Laue orientation, etc.? Actually, the questions to ask should be: (i) which orientation definition is the most relevant regarding the physical properties of interest? (ii) what are we able to measure, especially by means of neutron diffraction techniques?

In this chapter, we will only deal with the statistical description of the crystallographic texture, since it is the most relevant description regarding (i) the observation scale in neutron diffraction; and (ii) the macroscopic properties of a material.

2. Crystal symmetry and sample symmetry

2.1 Crystal symmetry: Definition of an orientation

Symmetry elements of the macroscopic properties of a crystal are restricted to proper and improper rotation operators. Translation operators have no influence here because the relevant parameter is the direction along which a property is measured: a translation operation does not affect this direction. Rotation (proper and improper) operators make a subgroup of the crystal space group which is called the point group: it is merely obtained from the space group by removing all translations and by moving all rotation axes through the same point. Point groups are therefore the relevant symmetry groups for discussing crystal macroscopic properties: the well-known Neumann's principle already stated in 1885 that "*the symmetry elements of any physical property of a crystal must include the symmetry elements of the point group of the crystal.*"

However, when performing a diffraction experiment, Friedel law implies centrosymmetry of diffracted intensities, even if the considered crystallographic structure is not centrosymmetric. This is the reason why the 32 point groups are gathered in 11 sets, called Laue classes which are defined by the 11 centrosymmetric point groups. They can be easily obtained by adding an inversion center to non-centrosymmetric point groups. Therefore, all crystal symmetries within a Laue class have the same influence on the symmetry of a Laue diffraction pattern. An "orientation" being defined as a transformation relation between two coordinate frames (for instance the goniometer frame and the crystal lattice frame), this major point means that, as far as diffraction intensities are considered (e.g., in pole figure measurements by diffraction methods, see Sect. 4), only Laue orientations

DOI: 10.1051/epjconf/201715500005

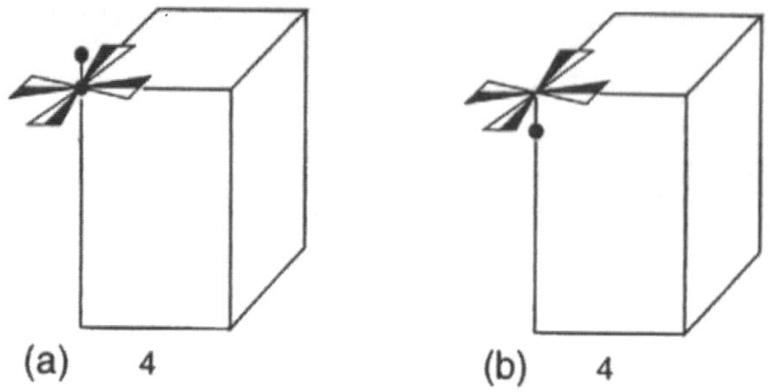

Figure 2. Example of two fictious enantiomorphs with artificial motifs, represented with the same *lattice* orientation. The crystal system is tetragonal. Crystal orientation (b) can not be obtained from (a) by applying an inversion center: (a) and (b) correspond to two different structures. However (a) and (b) can not be distinguished by means of "classical" diffraction methods since the latter artificially add an inversion center: (a) and (b) belong to the same Laue group (4), and have the same Laue orientation. (Figure reproduced from [1]).

(i.e. implying a center of symmetry which may or not be present in the crystal structure) can be distinguished. This has important consequences for materials which crystallize with a non-centrosymmetric point group, since diffraction artificially adds an inversion center. For instance, two enantiomorphs of the same material with the same lattice orientations but whose crystal orientations are related by a simple inversion have the same Laue orientation, and are impossible to distinguish by "classical" diffraction methods (Fig. 2). Another consequence is that, in the general case, only centrosymmetric physical properties can be analyzed, discussed, predicted or modeled from the single knowledge of the Laue orientation: non-centrosymmetric properties, described by odd-rank tensors (such as electric polarization, piezo-electricity, etc.), can be present in non-centrosymmetric crystals only, and their treatment thus requires the knowledge of the true structure orientation [2, 4].

In the following, the term "orientation" will always stand for "Laue orientation".

2.2 Sample symmetry

In practice, the shape of a polycrystalline material reflects its thermo-mechanical history: e.g., a plate is generally obtained by means of rolling, a wire by drawing, etc. A "sample" symmetry can thus be defined as a statistical symmetry resulting from its elaboration process, and should describe relationships between observed crystallites orientations. As a rule, this sample symmetry reflects the lowest symmetry deformation.

To fix ideas, let us consider for instance a rolling deformation process (Fig. 3). It involves a plane strain deformation of the polycrystal, and the resulting sample symmetry should be orthorhombic: if one considers a reference orthogonal frame made of the rolling direction (RD), transverse direction (TD) and normal (to the plate surface) direction (ND), plastic deformation during a rolling process induces grain rotations with mirrors symmetries with respect to (RD, ND), (TD, ND) and (RD, TD) planes. Hence, the sample symmetry is *mmm*: e.g., from a statistical point of view, as much grains with a lattice direction oriented at an α angle from RD are expected as grains with the same lattice direction at $-\alpha$ from RD. To provide another example, a torsion process, which involves a simple shear, should lead to a monoclinic sample symmetry.

This sample, statistical symmetry has important consequences on the description and representation of crystallographic textures, as will be detailed below (Sect. 3.2), and, of course, on the macroscopic physical properties: *the symmetry elements of any physical*

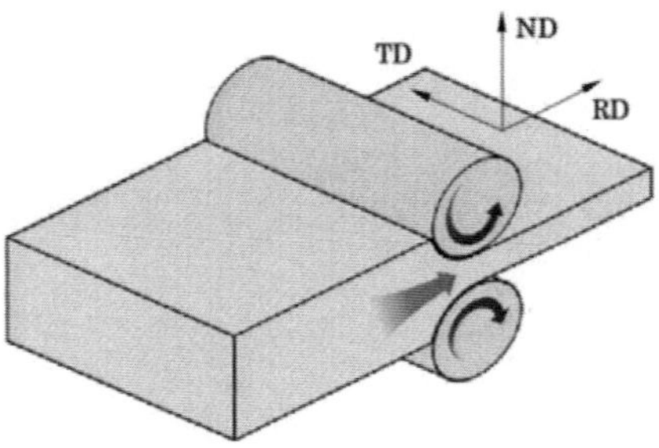

Figure 3. Rolling deformation process of a thick plate.

property of a polycrystalline sample must include the statistical elements common to its crystallographic texture [1]. The symmetry of a polycrystalline sample is generally reported by using the "*crystal symmetry – sample symmetry*" doublet notation. For instance, a rolled iron sheet generally exhibits a *cubic-orthorhombic* symmetry.

Finally, it is important to stress that, in a polycrystal, anisotropy of a physical property results from the anisotropy of this property in the single crystal *and* the presence of crystallographic texture, simultaneously: an isotropic single crystal property always implies an isotropic polycrystal property (even if the sample is highly textured), and the absence of texture always implies isotropy of macroscopic properties (even for highly anisotropic single crystal properties).

3. Representations of orientations – the orientation distribution function

3.1 Representations – Euler angles

In crystallographic textures analysis, representing an orientation consists in finding a description of the orientation of the crystallographic axes of a given crystallite, or a group of crystallites, within the polycrystal with respect to the chosen sample coordinate system, and vice-versa. Let us recall that it is always possible to define orthogonal coordinate systems for the crystal and the sample, whatever the crystal and sample symmetries: for instance, considering a general crystal lattice defined by **a**, **b** and **c** unit vectors (and α, β and γ lattice angles) the easiest way consists in defining a **z** vector normal to the (**a**, **b**) plane, a **y** vector parallel to **b**, and a **x** vector normal to the resulting (**y**, **z**) plane. The same kind of procedure can be applied to obtain an orthogonal (**X**, **Y**, **Z**) sample coordinate system.

There are various ways of describing an orientation. A classic, very simple way of describing an orientation, which is still widely used e.g., in metallurgy due to its compactness and simplicity, consists in making use of the Miller indices of two directions: the first $\{hkl\}$ indices correspond to the crystal plane whose normal coincides with the "**Z**" sample frame axis (e.g., the normal direction ND in the case of a rolled sheet), and the second $<uvw>$ indices correspond to the crystal direction which coincides with the "**X**" sample frame axis (e.g., the rolling direction RD). This leads to the well-known Miller notation $\{hkl\}$ $<uvw>$ for textures components (also called sometimes the *metallurgical representation*), which still allows the description of textures by means of "ideal orientations".

The most rigorous applied approach for representing orientations makes use of Euler angles and frame rotations. Use of orthogonal coordinate systems, as mentioned above, greatly simplifies the definition of Euler angles and of rotation matrices. However, there are several ways of defining Euler angles and corresponding sequence of rotations, leading to

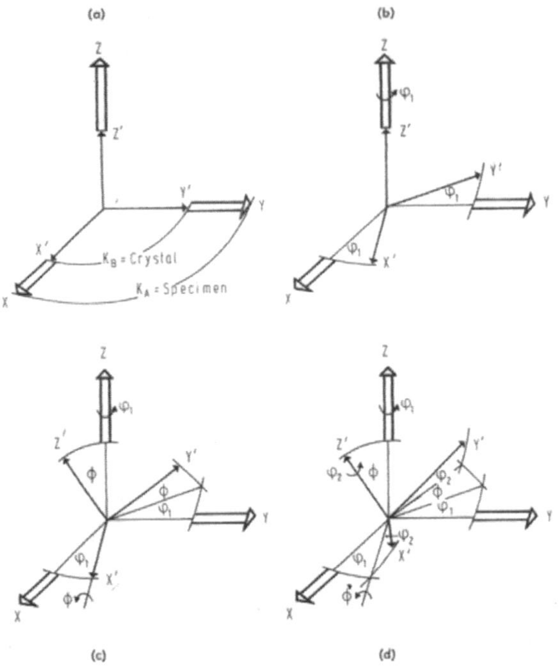

Figure 4. Bunge definition of Euler angles φ_1, Φ, φ_2 (see text). Here, the sample axis system is (X, Y, Z) and the rotated crystal axis system is (X', Y', Z') (from [7]).

various conventions (Kocks [5, 6], Bunge [7] and Roe [8], mainly). All these conventions are based on the fact that three angles are required to describe a crystal orientation, and hence they all lead to the definition of a three dimensional orientation space (the "Euler space"). Here, only the Bunge convention, which is likely to be the most frequently encountered one, will be detailed.

Bunge convention [7] makes use of the classical definition of Euler angles. To represent the orientation of the crystal axis system (CAS) with respect to the sample axis system (SAS), which is taken as the reference frame, three successive intrinsic rotations are considered, starting with the crystal axes overlapping the sample axes (Fig. 4). First, the CAS rotates by φ_1 about the **Z** axis. Secondly, the CAS rotates by Φ about the now rotated **x** axis. Finally, the CAS rotates by φ_2 about the new **z** axis. Each of these three elemental rotations can be represented by a rotation matrix, and the CAS orientation can thus be represented by the product of these three elemental rotation matrices. The transformation matrix describing the rotations from the SAS to the CAS, i.e., expressing the crystal lattice unit vectors as a function of the sample unit vectors (*active rotation* convention) is thus written as:

$$\mathbf{g} = \begin{pmatrix} \cos\varphi_1\cos\varphi_2 - \cos\Phi\sin\varphi_1\sin\varphi_2 & -\cos\varphi_1\sin\varphi_2 - \cos\Phi\cos\varphi_2\cos\varphi_1 & \sin\varphi_1\sin\Phi \\ \cos\varphi_2\sin\varphi_1 + \cos\varphi_1\cos\Phi\sin\varphi_2 & \cos\varphi_1\cos\Phi\cos\varphi_2 - \sin\varphi_1\sin\varphi_2 & -\cos\varphi_1\sin\Phi \\ \sin\Phi\sin\varphi_2 & \cos\varphi_2\sin\Phi & \cos\Phi \end{pmatrix}. \tag{1}$$

In the following, an orientation will be represented by its corresponding **g** matrix.

It is to note that other approaches can be found in literature for representing orientations (or misorientations between grains), such as the axis/angle description, the Rodrigues vector description (which combines a rotation axis and a rotation angle into an appropriate vector in a so-called Rodrigues space) [9], or the quaternion description (in a 4-dimensional Euclidian

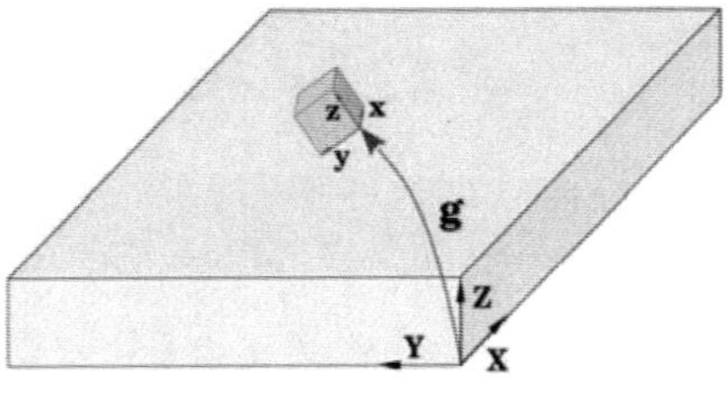

Figure 5. Definition of a **g** orientation.

space) [10]. These specific approaches have advantages and disadvantages over Euler angles approaches for specific applications, but these will not be treated here.

3.2 The orientation distribution function

With the description of orientations by Euler angles, each crystallite c, characterized by its own orientation $\mathbf{g_c}$, corresponds to one point whose coordinates are $(\varphi_1, \Phi, \varphi_2)_c$ in the three-dimensional orientation space. Considering all the crystallites which constitute the polycrystalline aggregate thus allows defining the distribution of points in the orientation space, namely the Orientation Distribution (OD). The OD fully and quantitatively describes the crystallographic texture, which is then taken as a set of discrete orientations. However, it is a global, statistical description, at the scale of the polycrystalline aggregate (or a small part of it), ignoring any possible correlation between a grain orientation and its position within the polycrystal. As a consequence, OD provides no information about the local environment of each grain, which may be another important parameter for property calculations: this kind of description (not treated here) is associated to the definition of "orientation correlation functions", whose determination requires local, microscopy techniques.

For practical purposes and easy manipulation, OD is generally mathematically described by a continuous function, the Orientation Distribution Function $f(\mathbf{g})$, which has to be fitted to the data, and defined as:

$$f(\mathbf{g})\mathrm{d}\mathbf{g} = \frac{\mathrm{d}V(\mathbf{g})}{V}, \tag{2}$$

i.e., $f(\mathbf{g})$ is the volume fraction of grains characterized by an orientation within a region $\mathbf{dg}$ around $\mathbf{g}$ orientation. With this definition, the ODF may be assimilated to a probability density, but it is generally normalized so that for a fully randomly oriented material, with no crystallographic texture, $f(\mathbf{g}) = 1\ \forall\mathbf{g}$, hence:

$$\frac{1}{8\pi^2}\iiint f(\varphi_1, \Phi, \varphi_2)\sin\Phi \mathrm{d}\varphi_1 \mathrm{d}\Phi \mathrm{d}\varphi_2 = 1 \tag{3}$$

the orientation space volume being equal to $8\pi^2$. The infinitesimal volume element in the orientation space is thus $\frac{1}{8\pi^2}\sin\Phi\mathrm{d}\varphi_1\mathrm{d}\Phi\mathrm{d}\varphi_2$.

The ODF allows the quantitative description of the crystallographic texture, namely the calculations of volume fractions of crystallites characterized by a $\mathbf{g}$ orientation, with a tolerated $\mathbf{\Delta g}$ misorientation: such a set of grains is called a *texture component*. Practically, textures are often described in terms of "ideal orientations": some which are frequently observed[2] have specific names, such as Cube (corresponding to {100}<010> orientations in

[2] E.g., after some particular thermo-mechanical treatment.

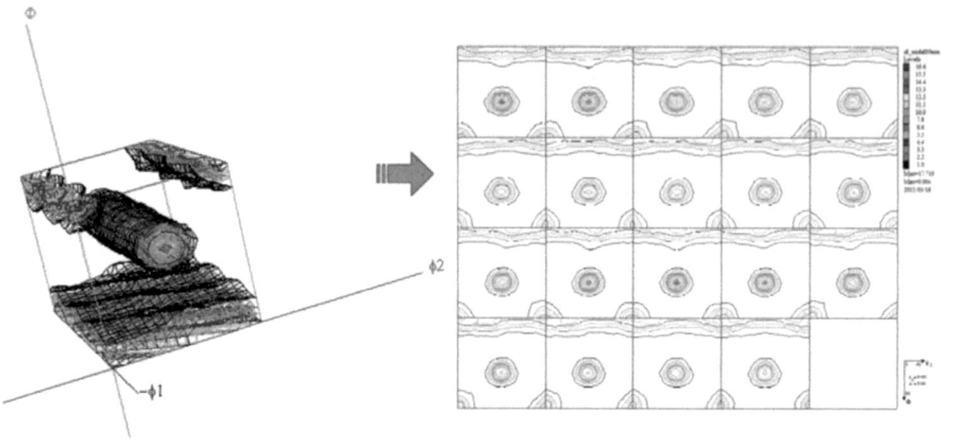

Figure 6. Example of an ODF obtained on a wire-drawn Al alloy, evidencing a {*hkl*} <*111*> fiber texture component (parallel to φ_1 axis): (left) 3D representation; (right) 2D slices at constant φ_1.

Miller notation), Copper ({211} <111>), Goss ({110} <001>), etc... for the cubic crystal symmetry. Some textures are also better described as *fiber* textures (or in term of *fiber* components) when a rotational symmetry with respect to a sample direction is present: for instance, α {*hkl*} <*110*> and γ {*111*} <*uvw*> are typical fiber texture components observed in rolled bcc metals, and correspond to rotational symmetries with respect to RD or ND, respectively.

The ODF can be represented in the three-dimensional orientation space by taking the three Euler angles as Cartesian coordinates, with orthogonal axes. From the Bunge's definition of the Euler angles, this orientation space is periodic in all three axis directions, with a 2π period, thus defining a unit cell ($\varphi_1 \in [0, 2\pi]$, $\Phi \in [0, 2\pi]$, $\varphi_2 \in [0, 2\pi]$). Most of the time, for practical purposes, the ODF is plotted on two dimensional slices of the orientation space, selected at constant $\Delta\varphi_1$, $\Delta\Phi$ or $\Delta\varphi_2$ steps (Fig. 6). One dimensional sections may be also used in the case of fiber textures, since in that case the ODF is a function of only two parameters.

Considering the definition of $\mathbf{g}$ matrix in (1), $\mathbf{O_C}$ being a crystal symmetry operator and $\mathbf{O_S}$ a sample symmetry operator, an equivalent orientation $\mathbf{g}'$ is obtained by:

$$\mathbf{g}' = \mathbf{O_C}\,\mathbf{g}\,\mathbf{O_S}. \tag{4}$$

These $\mathbf{g}$ and $\mathbf{g}'$ orientations correspond to different points in the orientation space, but these two corresponding rotation matrices represent indistinguishable objects: therefore the ODF must be invariant with respect to the symmetry operations:

$$f(\mathbf{g}') = f(\mathbf{O_C}\,\mathbf{g}\,\mathbf{O_S}) = f(\mathbf{g}). \tag{5}$$

A fundamental zone (or an asymmetric unit) can thus be defined in the orientation space so that each orientation is uniquely described by a single point. Its size depends on the number of crystal and sample symmetry operators: the total volume of orientation space should be divided by the number of symmetry elements. Unfortunately, in the Euler angles representations, this fundamental zone generally has curved boundaries[3]: practically, a reduced orthogonal orientation space is thus defined from the classical Cartesian, orthogonal

[3] This is not the case in the Rodrigues representation.

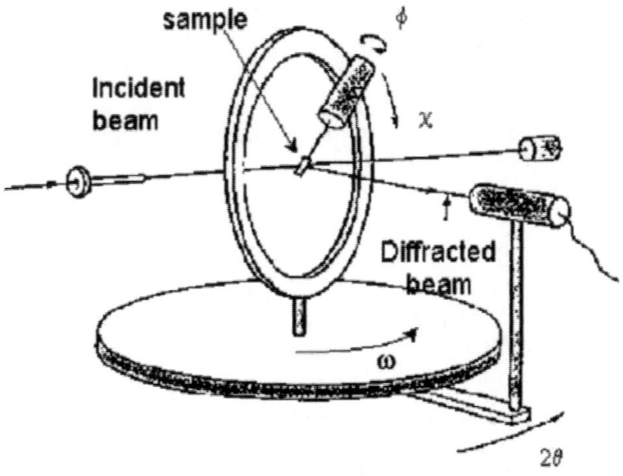

Figure 7. Principle of a 4-circles diffractometer, and definition of rotation angles.

Euler space so that the minimum number of equivalent orientations are contained within appropriate ranges of Euler angles. For instance, for the cubic-orthorhombic symmetry, a reduced, cubic region ($\varphi_1 \in [0, \pi/2], \Phi \in [0, \pi/2], \varphi_2 \in [0, \pi/2]$), containing three copies of the actual fundamental zone is used.

Texture index is a widely used parameter globally reflecting the crystallographic texture of a material over the whole orientation space. It is obtained from the ODF knowledge by:

$$I = \int f^2(\mathbf{g})\mathrm{d}\mathbf{g}. \tag{6}$$

It is equal to 1 for a purely random orientation distribution, and increases with the texture *acuteness*, i.e., the reinforcement of preferential orientations.

4. Diffraction methods

4.1 Pole figures measurements

The OD (or the ODF) can not be directly determined by an experiment. X-ray and neutron diffraction methods for measuring textures are based on the fact that the diffracted integrated intensity for a specific *hkl* reflection and for a specific orientation of the diffraction vector (with respect to the sample) is proportional to the diffracting volume, namely the volume of crystallites which are correctly oriented to satisfy the Bragg condition for *hkl*.

Experimentally, "measuring" a texture by diffraction thus merely consists in measuring the variations of the diffracted intensity with respect to the diffraction vector $\mathbf{Q}_{hkl}$ orientation. Using a monochromatic X-ray or neutron source, a classical instrument to reach that purpose is the 4-circles diffractometer (Fig. 7): the polycrystalline sample is mounted at the center of an Eulerian cradle, then the diffracted intensity is recorded as a function of the (χ, ϕ) sample orientation, for various *hkl* reflections, in $(\theta, 2\theta)$ geometry. As a rule, low index *hkl* reflections are measured in priority since their intensities are usually the highest and they are less subject to peak overlaps. As a result, raw (*hkl*) diffraction pole figures are obtained, which are projection maps of the diffracted intensities. After experimental corrections and a normalization process, these provide experimental *direct* pole figures.

A direct *hkl* pole figure (PF) is the stereographic projection on the sample (**X**, **Y**) plane of the distribution of (*hkl*) crystal "poles", a pole being defined as the normal direction to a

 DOI: 10.1051/epjconf/201715500005

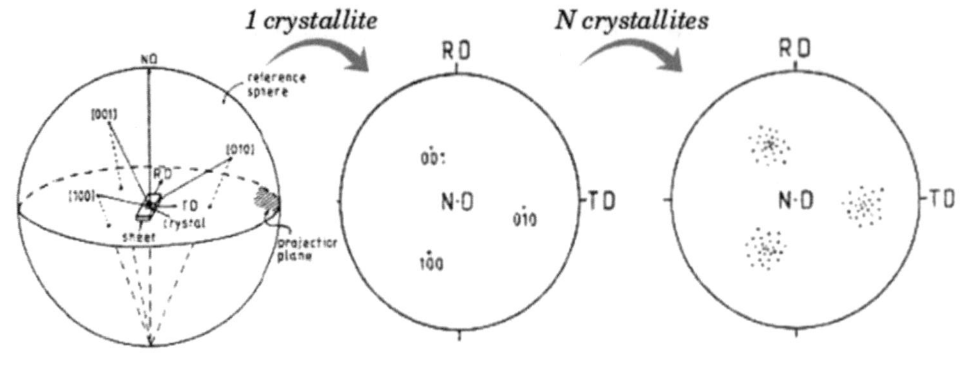

Figure 8. Principle of the construction of a 100 pole figure [7].

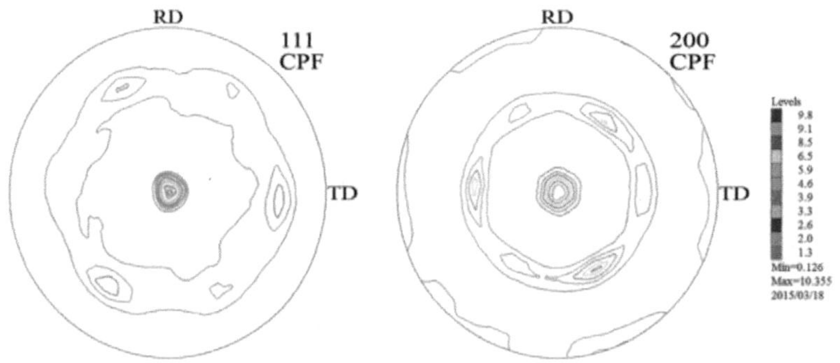

Figure 9. Example of 111 and 002 experimental pole figures measured on a wire-drawn Al alloy (ODF shown in Fig. 6 was calculated from these PFs).

(*hkl*) crystallographic plane. This direction is always taken centrosymmetric: a single crystal Bragg reflection is associated to a pole. Of course, for a given family of equivalent $\{hkl\}$ crystallographic planes, each crystal constituting the polycrystalline aggregate contributes symmetrically equivalent poles (Fig. 8). Plotting all the poles for all crystallites allows to deduce the pole density $p_{hkl}(\chi, \phi)$ for the polycrystal (Fig. 9). Normalization of experimental pole figures consists in imposing a pole density value equal to 1 for all (χ, ϕ) orientations in the case of a random texture, and in making use of the fact that the total diffracted intensity over a sphere (i.e., over a 4π solid angle) is independent of the crystallographic texture, hence:

$$\int p_{hkl}(\chi, \phi) \sin \chi \mathrm{d}\chi \mathrm{d}\phi = 4\pi. \tag{7}$$

As a result, a normalized experimental *hkl* pole figure is the polar plot of $p_{hkl}(\chi, \phi)$, namely the volume fractions of the sample for which the (*hkl*) plane normal has a (χ, ϕ) orientation in the sample coordinate system.

Direct pole figures can be used to characterize qualitatively the crystallographic texture. They are generally useful to identify typical, "well-known" texture components. Moreover, it is noteworthy that pole figures directly reflect the sample symmetry, and so the symmetry of a deformation process: in some favorable cases, the latter may even be identified by merely observing the symmetries of pole figures (an example is given in Sect. 5). But to quantitatively characterize a texture, the Orientation Distribution Function has to be computed from a set of Pole Figures.

4.2 From pole figures to the orientation distribution function

Basically, a normalized pole figure for $\{hkl\}$ is obtained by integration of the orientation distribution function along paths corresponding to 2π rotations of crystallites about their corresponding *(hkl)* poles. This is traduced in the fundamental equation:

$$p_{hkl}(\chi, \phi) = \frac{1}{2\pi} \int_0^{2\pi} f(\varphi_1, \Phi, \varphi_2) \mathrm{d}\Gamma \tag{8}$$

where Γ corresponds to the rotation angle with respect to the *(hkl)* diffracting plane normal. It is to note that, excepting special cases such as *(00l)*-type poles, the integration path is generally not a straight line in the orientation space. A pole figure can thus be considered as a bi-dimensional projection of the (three-dimensional) ODF. Computing any PF when knowing the ODF is a trivial task, but computing the ODF with the single knowledge of few (typically between 2 and 5) PFs is more difficult: it may be compared with the tomography problem of reconstructing a three-dimensional volume from bi-dimensional images, but with an extremely reduced available amount of images!

Two classes of methods are available for computing ODF from PFs: the harmonic methods, and the discrete (or direct) methods. Both are briefly described in the following, as well as their respective advantages and drawbacks.

4.2.1 Harmonic methods [8,11]

The ODF is periodical with respect to the three rotations characterized by φ_1, Φ and φ_2 Euler angles, whereas a pole figure is periodical with respect to the two rotations characterized by χ and ϕ angles (in our convention). On the one side, the ODF can thus be expanded in a series of generalized spherical harmonics, which are orthogonal functions similar to the "classical" spherical harmonics functions but considering three rotations instead of two:

$$f(\varphi_1, \Phi, \varphi_2) = \sum_{l=0}^{\infty} \sum_{m=-l}^{+l} \sum_{n=-l}^{+l} C_l^{mn} T_l^{mn}(\varphi_1, \Phi, \varphi_2), \tag{9}$$

where C_l^{mn} are the ODF coefficients (to be determined), and T_l^{mn} are the generalized spherical harmonic functions defined by:

$$T_l^{mn}(\varphi_1, \Phi, \varphi_2) = e^{im\varphi_2} P_l^{mn}(\cos \Phi) \, e^{in\varphi_1}, \tag{10}$$

the P_l^{mn} being the generalized (and normalized) Legendre polynomials.

On the other side, each pole figure can be expanded in a series of spherical harmonic functions:

$$p(\chi, \phi) = \sum_{l=0}^{\infty} \sum_{m=-l}^{+l} Q_l^m Y_l^m(\chi, \phi), \tag{11}$$

where Q_l^m are the corresponding pole figure coefficients, which can be "easily" calculated from experimental data, and Y_l^m are the spherical harmonic functions defined by:

$$Y_l^m(\chi, \phi) = P_l^m(\cos \chi) \, e^{im\phi}, \tag{12}$$

the P_l^m being the Legendre polynomials.

Crystal and sample symmetries induce selection rules on the C_l^{mn} coefficients: some may cancel, or some relationships may be obtained between them, thus simplifying the expansions. But the relationships between the C_l^{mn} coefficients can become complicated to determine for high symmetries: this problem was treated in an elegant way by Bunge [7], whose formalism makes use of *symmetrized* spherical harmonic functions which are expressed as linear combinations of the T_l^{mn} and Y_l^m functions, respectively, so that invariance with respect to symmetry operations is verified.

As a result from previous considerations and Eq. (8), a set of linear equations relating the PF coefficients and the (unknown) ODF coefficients is obtained. This set can be solved if the number of C_l^{mn} unknowns is smaller than or equal to the number of equations, which depends on the number N of available independent experimental PFs: as a result, N and the crystal symmetry both fix the maximum degree $l_{\max}$ of the expansion (9). For instance, for cubic crystal symmetries, with only two pole figures, the C_l^{mn} coefficients can be determined up to $l_{\max} = 22$, but for triclinic crystal symmetries, 29 independent pole figures would be required to determine the C_l^{mn} coefficients only up to $l_{\max} = 14$! Practically, the C_l^{mn} are obtained by minimizing the square deviation between the experimental and the re-calculated (expanded in SH series) pole figures, imposing $p_{hkl}(\chi, \phi) \geq 0$ and $f(\varphi_1, \Phi, \varphi_2) \geq 0$ (positivity constraints).

Harmonic methods allow fast computations and low computing memories since each order of the expansion can be computed separately. Moreover, as a result, a compact representation of the ODF is obtained, since only the ODF coefficients need to be known to reconstruct the whole ODF: this allows an easy manipulation in property calculations.

However, as above-mentioned, the required number of PF can become very large (and sometimes unrealistic) for low crystal symmetries. Furthermore, whatever the symmetry degree, the expansion in a series must be practically truncated at a certain order, and this truncation should not drastically affect the results: these methods thus implicitly assume that the ODF should be a smooth function, i.e., that the texture is not too sharp, which may be too restrictive in some cases. Besides, "*ghosts*" peaks may be obtained in the computed ODF. Indeed, by definition, a pole distribution on a sphere is always centrosymmetric: as a consequence of Friedel's law, *hkl* and $\overline{hkl}$ pole figures are always superposed. But spherical harmonic functions (and the generalized harmonic functions) are centrosymmetric when l is even only, and antisymmetric otherwise. This means that the C_l^{mn} coefficients with an odd l can not be determined by "classical" diffraction. Hence the ODF is generally written as the sum of an even part, which can be determined from experimental data, and an odd part which has no effect on PF but, if neglected, may induce artefacts and spurious peaks (named "*ghost peaks*") in the ODF. Several methods have been developed to deal with this issue which is especially critical in non-centrosymmetric crystal symmetries [12–16].

4.2.2 Discrete methods

Here, the ODF computation is directly performed in the orientation space. These methods are based on the representation of PFs and ODF by sets of discrete values, obtained by defining regular grids in the projection space (the pole figures) and in the orientation space, respectively. Projection paths relating ODF cells with PFs cells have then to be established. Of course, each PF cell corresponds to the projection of several ODF cells, and the projection lines depend on the crystal and sample symmetries. As a result, the Eq. (8) can be rewritten in its discrete form:

$$p_{(hkl)}(\mathbf{y}) = \frac{1}{N}\sum_{i=1}^{N} f(\mathbf{y} \Leftarrow \mathbf{g}_i) \tag{13}$$

the arrow symbolizing the projection path to the $\mathbf{y}$ cell in the pole figure from the N orientation cells $\mathbf{g}_i$ which contribute to the pole density in $\mathbf{y}$. The resolution of the resulting system of linear equations thus requires the inversion of a large matrix which describes all the projection paths, and the solution is likely not to be unique if not enough experimental pole figures are available.

Various algorithms for implementing discrete methods have been developed during the last decades. For instance, the Vector Method [17] performs direct inversion of the matrix by introducing geometry constraints on the grids. Other methods, such as WIMV [18] or ADC [19] algorithms, first make an initial estimate of the ODF which is used to compute the corresponding initial pole figures. The latter are compared with the experimental pole figures, then correction factors are applied to the ODF to compute new pole figures, and the process is repeated until a good accordance between the calculated and the experimental pole figures is obtained.

Discrete methods have many advantages over harmonic methods. They are generally better suited to sharp textures, but also to low symmetries, since the required amount of pole figures may be reduced. Moreover, most of softwares which implement discrete methods easily allow to impose some positivity and zero-range constraints on the ODF, and also to take into account various symmetries. However, these methods require lots of computing memory, since the matrix relating the orientation space and the pole figures can be very large (although sparse): efficient and optimized programming is thus highly required. Experimental data accuracy is also more critical than with harmonic methods. Besides, the obtained discrete ODF, which consists in a large set of ($\mathbf{g}$, $f(\mathbf{g})$) doublets, is generally less fit for use in e.g., property calculations than reduced sets of coefficients as obtained by means of harmonic methods. Finally, let us mention that, in some cases, if additional constraints are not introduced, ghosts peaks may also appear in the obtained ODF, due to rank deficiency in Eq. (13).

4.3 Neutron diffraction vs. X-ray diffraction

4.3.1 X-ray diffraction

X-ray diffraction is the most widely used technique for measuring pole figures because it is generally fast and easy to perform on X-ray laboratory diffractometer. Most of the time, the latter generally consists in a 4-circles diffractometer equipped with an Eulerian cradle, quite similar to those used for single crystal X-ray diffraction (cf. Sect. 4.1), with Bragg-Brentano focussing and $\theta - 2\theta$ geometry. Due to the weak penetration depth of X-rays in most materials (e.g., few μm or tens of μm in metallic alloys), most measurements are performed in reflexion mode, since transmission mode measurements would require thin samples. So most X-ray measurements are *surface* measurements: for instance, when measuring an iron sample in reflexion mode using Mo$K\alpha$ radiation, 90% of the diffracted intensity comes from a layer only $\sim 10\mu$m thick below the surface when 2θ angle is close to 40° [20]. Moreover, when the angle between the incident X-ray beam and the sample surface decreases, the shape of the irradiated area tends to an increasingly elongated ellipse and the interaction volume increases as well, leading to a decrease of the diffracted signal that reaches the detector (i.e., that passes through the detector slits within a fixed solid angle): this is the *defocussing effect*, which requires important corrections to be performed on the measured intensities, and practically prevents any measurement when the tilt angle is below $\sim 25°$. As a consequence, only incomplete pole figures can be obtained by X-ray diffraction, corresponding to the central area of the polar plot in reflexion mode, i.e., to the low χ values only (Fig. 10). This is an important issue for computing ODF, especially applying harmonic methods, and the missing part of the pole figures generally has to be extrapolated. To experimentally obtain complete pole figures, the two configurations (and so two series of measurement) are required : reflexion mode provides the low χ region of the PFs, and

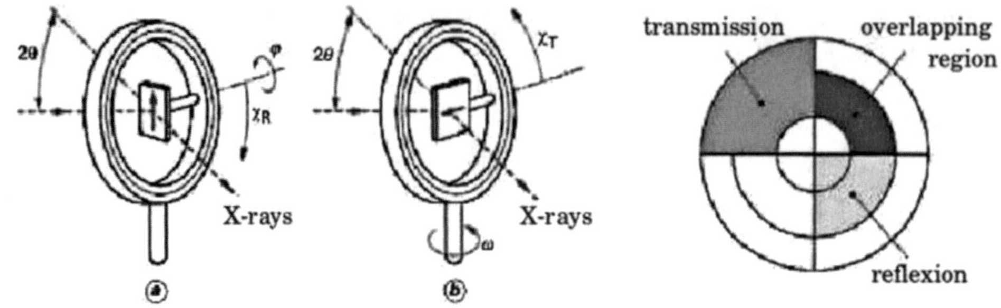

Figure 10. Left: the two pole figure measurement modes in X-ray diffraction: (a) reflexion; (b) transmission; Right: combination of pole figure regions obtained for each mode (figure reproduced from [21]).

transmission mode (when possible) provides the high χ region. Then the two sets of data have to be properly merged (Fig. 10).

Besides, a diffraction pole figure should plot the diffracted intensity variations due solely to the different volumes of grains in diffraction condition as a function of the sample orientation. The intensity variations due to the different X-ray path lengths within the matter as a function of the sample orientation have thus to be properly corrected. But the X-ray absorption coefficients are large for most elements, and small variations in these path lengths may induce quite large intensity variations. As a consequence, absorption correction is another critical step in the X-ray data treatment.

Rq. It is to note that crystallographic texture measurements can also be performed by means of X-ray synchrotron radiation diffraction. However, it is most of the time difficult to obtain diffraction patterns for numerous sample orientations. Using a bi-dimensional detector, and when the number of available Debye rings is large, it is possible in some specific cases to determine the orientation distribution with a single sample orientation [22]. Taking advantage of the small beam sizes, synchrotron diffraction is better suited to the mapping of orientations at the grain scale, e.g., by means of 3D-XRD [23], or Laue micro-diffraction [24] techniques. Obtained information is then generally far from being statistically representative of the whole sample, and these local measurement techniques are not used to determine orientation distributions.

4.3.2 Neutron diffraction

The main advantage of neutrons over X-rays is their low absorption by most elements [25]. Measurements can thus be performed in transmission mode on massive samples to obtain complete pole figures in a single experiment. The only necessary data correction to be carried out is the background subtraction: absorption correction is generally useless, and there is no defocussing effect. Hence data processing is greatly simplified. Sample volumes of the order of a cm^3 can be characterized, allowing texture analysis over statistically representative amounts of crystallites, and, as a consequence, errors on the computed ODF are generally significantly lower than with X-ray diffraction: the mean difference between re-calculated and experimental pole figures can be smaller than $\sim$5%, to be compared with $\sim$15% classically obtained from XRD data.

The ideal sample geometry is the sphere, since absorption is then independent on the sample orientation. But due to the low neutron absorption coefficients for most materials, there is generally a quite large tolerance on the sample geometry without having to correct from anisotropic absorption effects: for instance, between two Fe thicknesses of 5 mm and 2 mm, the absorption difference of thermal neutrons is close to 6% only, which practically induces a negligible effect on the measured diffracted intensities. Most samples may thus have cubic-like shapes (much easier to obtain than spheres), or even cylindrical or tubular shapes, or still more irregular shapes. Moreover, no sample surface preparation is required.

As a consequence, neutron diffraction for crystallographic texture analysis is more specifically a tool of choice in the following cases:

- coarse-grained materials (with grain size $\sim$ mm^3);
- multiphase materials (especially with anisotropic arrangement of the phases, and resulting anisotropic absorption which may be a strong handicap for X-ray measurements);
- geological materials;
- minority phases characterization (down to volume fractions $\sim 1\%$);
- use of complex sample environments, such as furnaces, cryostats, mechanical testing machines, etc: a typical application consists in in situ monitoring texture evolutions during thermal treatments;
- low crystal symmetries (since neutron scattering lengths are independent of $\sin\theta/\lambda$, high index reflections may be measured with a higher accuracy than with X-ray diffraction);
- magnetic texture analysis.

4.4 Neutron diffractometers

4.4.1 Continuous sources

On continuous neutron sources, crystallographic texture measurements are generally performed with a monochromatic neutron beam, and the typical dedicated instrument is a 4-circles diffractometer (Fig. 7), as previously mentioned in Sect. 4.1. The sample is mounted at the center of an Eulerian cradle which allows to impose any orientation of the diffraction vector with respect to the sample coordinate system thanks to the two rotations defined by the χ and ϕ angles.

When the detector consists in a single counter, the diffraction geometry is $\theta - 2\theta$, so during a pole figure measurement, at a constant 2θ detector position, the ω angle is fixed so that $\omega = \theta$. In that case, for a selected *hkl* pole figure, 2θ is the detector position corresponding to the maximum intensity of the *hkl* diffraction peak: only the *peak* intensity is recorded, not the *integrated* intensity (which would require time-consuming $\theta - 2\theta$ scans for each (χ, ϕ) orientation). This implies that the position of the *hkl* peak maximum should not significantly vary with (χ, ϕ), e.g., due to internal strains or to a misalignment of the sample: using sufficiently large beam section and divergence (leading to broadened diffraction peaks) generally prevents such effects. However, it is noteworthy that in the case of low symmetry or multiphase materials, diffraction peaks are likely to overlap, and beam divergence should be properly reduced to allow a good peak separation. Pole figures are recorded by scanning χ and ϕ rotations: a classical measurement grid is defined with $\Delta\chi = \Delta\phi = 5°$, leading to 1368 orientations to cover a whole pole figure. This may take between 3 and 12 h, depending on the neutron flux, the collimation, the *hkl* structure factor and the absorption. The main advantage is the great simplicity of the measurements and of the data treatment. A major limitation of this constant wavelength technique occurs for low symmetry materials whose texture analysis requires the measurement of numerous pole figures with frequent peak overlaps to deal with.

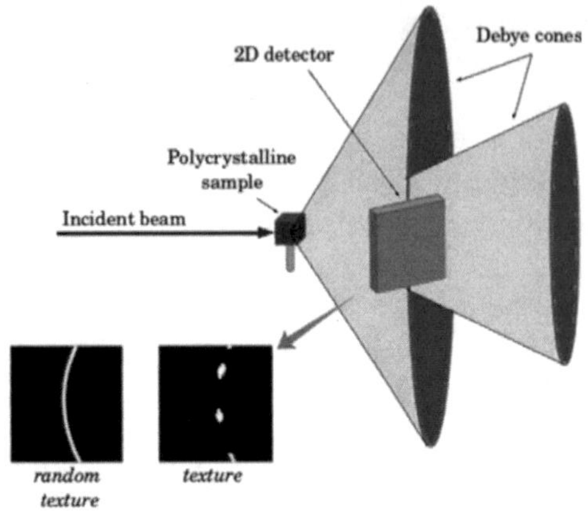

Figure 11. Intercept of a Debye cone by a bi-dimensional flat detector, and example of 2D diffraction images obtained in the case of a random texture (homogeneous Debye rings) and in the case of a textured sample (non homogeneous, "spotted" rings). Each pixel on the detector corresponds to a different **Q** orientation.

And of course the very low time resolution excludes time-resolved studies from the range of applicability of this technique.

Linear (1D) or bi-dimensional (2D) *position sensitive detectors* (PSD) can provide interesting alternatives to the single counters[4]. On the one hand, a linear PSD covering a wide 2θ range allows the simultaneous recording of a whole diffractogram, i.e., of numerous *hkl* Bragg peaks for a single sample orientation: by applying appropriate ω, χ, ϕ rotations to the sample, several *hkl* pole figures can thus be obtained in a reduced amount of time. This may be especially valuable in the case of in situ, time-resolved experiments, e.g., when characterizing recrystallization or phase transformations at high temperature. Moreover, full diffraction profiles are obtained, and integrated intensities can be used instead of peak intensities, generally leading to a better accuracy. Specific data treatment procedures using Rietveld analysis have been developed to extract texture information from sets of whole diffractograms [26–28].

On the other hand, bi-dimensional detectors can intercept a *hkl* Debye diffraction cone over a given azimuthal range (Fig. 11): several diffraction vector $\mathbf{Q}_{hkl}$ orientations are thus measured simultaneously, leading to the measurement of numerous (*hkl*) poles at the same sample orientation. Each pole figure measurements requires less sample rotations, and hence less measurement time. Again, integrated intensities can be extracted. However, data treatment is more complicated than when using a single counter, since it involves image processing, with background, flat-field, and geometric corrections, followed by azimuthal integrations of recorded intensities and *a posteriori* calculation of the "true" (χ, ϕ) orientation of the diffraction vector. Quite huge amount of data have to be stored and processed, requiring efficient storing and computing solutions.

For instance, 6T1 diffractometer located at LLB (CEA Saclay) [29] uses monochromatic thermal neutrons ($\lambda = 1.159$Å) selected by a Cu(111) monochromator at a fixed take-off angle ($2\theta_m = 32°$). An innovative Eulerian cradle with three integrated translations has been

[4] Rq. It is noteworthy that, when using a PSD, the diffraction geometry is not strictly $\theta - 2\theta$ anymore: orientation of the diffraction vector slightly moves within the 2θ range over which the diffraction peak is defined. The two polar angles defining the orientation of **Q** in the sample axis system are not merely (χ, ϕ) anymore as in $\theta - 2\theta$ geometry but have to be re-calculated from the knowledge of $(\omega, \chi, \phi, 2\theta)$ angles.

Figure 12. 6T1 diffractometer, at LLB, equipped with its bi-dimensional detector. A small tensile machine is mounted on the Eulerian cradle for in situ mechanical testing.

Figure 13. The 12 positions sample changer of 6T1 diffractometer.

recently mounted, allowing texture mapping or local measuremets in relatively massive samples (Fig. 12). Textures can be measured by using either a highly efficient single detector in $\theta - 2\theta$ geometry, or a bi-dimensional detector ($200 \times 200\,\text{mm}^2$, covering $\sim 12°$ in 2θ). A set of four radial collimators allows covering a wide range of divergences and analysis volumes. An original sample changer has been developed to sequentially measure textures of twelve samples without external manual operation (Fig. 13). A small tensile machine is also available for in situ measurements.

STRESS-SPEC diffractometer at FRM-II (Munich) [30] is equipped with three different monochromators (Ge(511), bent Si(400) and PG(002)), and the take-off angle can be continuously varied from $2\theta_m = 35°$ to $110°$, allowing to use wavelengths between 1 Å and 2.4 Å: a compromise between resolution and intensity can thus be found for each problem. A full circle Eulerian cradle is available for "light" samples, as well as a 1/4 circle cradle, for heavy samples. A 6-axis robotic arm can also be mounted, which offers more flexibility and can be used as a sample changer (Fig. 14). Detection is performed by a bi-dimensional detector ($300 \times 300\,\text{mm}^2$).

Finally, at ILL, let us mention D1B, the high flux 2-axis powder diffractometer with its linear PSD covering 128° in 2θ, and D20, the very high intensity 2-axis diffractometer with variable resolution, equipped with a large microstrip detector covering 153° [31]. These instruments are mostly dedicated to phase transitions, magnetic structures or kinetics studies,

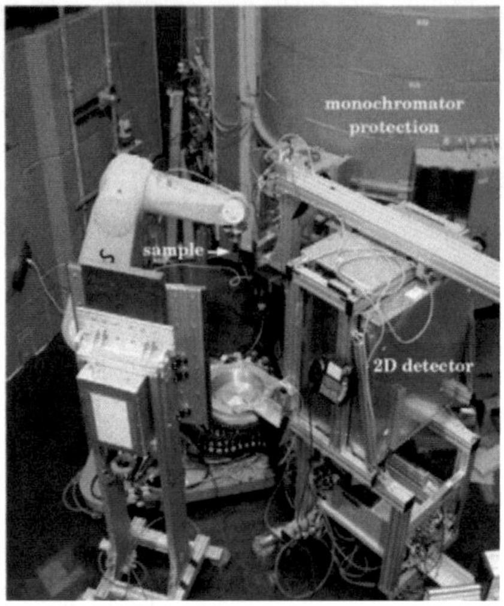

Figure 14. STRESS-SPEC diffractometer, at FRM-II, equipped with its robotic arm.

Figure 15. D1B diffractometer, at ILL, equipped with an Eulerian cradle. The "banana"-type multidetector is inside the green "box".

but they can also be equipped with an Eulerian cradle to perform crystallographic texture measurements (Fig. 15).

4.4.2 Pulsed sources

On pulsed neutron sources, a polychromatic beam is used, and a whole diffractogram $I = f(|\mathbf{Q}|)$ can be measured at a fixed detector position by using the *time of flight* (TOF) technique[5]. So by taking advantage both of TOF and of several counters or bi-dimensional detectors located at various angular positions all around the sample, several *hkl* pole figures can be measured simultaneously with very limited sample rotations. This is highly valuable for performing time-resolved studies, such as recrystallization or phase transformation studies, and/or for measurements involving complex sample environments,

[5] TOF may also be used on continuous (fission reactor) sources but a temporal (pulsed) structure has then to be applied to the white neutron beam by using choppers, leading to a drastic decrease of the neutron flux incoming the sample. On a pulsed spallation source, the neutron beam naturally exhibits a temporal structure well suited to TOF techniques.

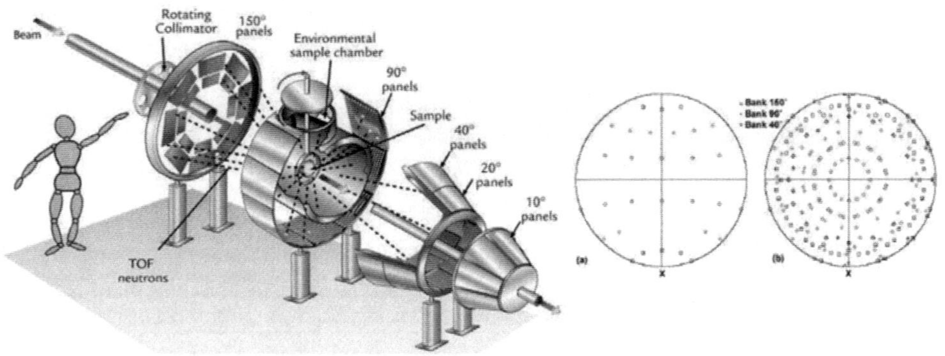

Figure 16. Left: sketch of HIPPO TOF diffractometer, at LANSCE, showing the arrangement of detector tubes in panels arranged on rings of constant diffraction angles. Right: covering of PFs by 3 detector banks for (a) 1 sample orientation and (b) 8 sample orientations [32].

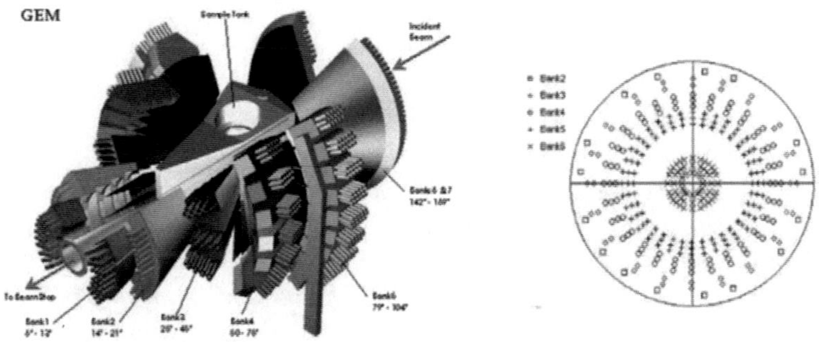

Figure 17. Left: sketch of GEM TOF diffractometer, at ISIS. Right: covering of PFs by 5 detector banks for 2 sample orientations [33].

such as furnaces, pressure cells, etc.. This is also extremely interesting for measuring textures non-destructively in relatively large, precious samples (e.g., archeological objects) which would not fit on an Eulerian cradle. TOF diffractometers are instruments which are by far more complex and expensive than monochromator-based diffractometers, especially due to the huge detector banks and their associated electronics. Besides, TOF data are slightly more difficult to process and analyze than constant wavelength data and data treatment takes naturally high benefit of full-pattern (Rietveld) approaches.

For instance, the detector system of the HIPPO diffractometer at LANSCE [32] consists in 1360 ^{3}He detector tubes, arranged in panels, covering a detector area of 4.8 m^2 and scattering angles from 10° to 150° (Fig. 16): only eight sample rotations provide a good coverage of pole figures, with a counting time close to 20 min (Fig. 16). GEM diffractometer at ISIS [33] is equipped with 7270 tubes arranged in 6 detector banks, covering a scattering angle range from 1° to 169°. For texture measurements, detectors are gathered in 164 groups, each covering $\sim 10° \times 10°$. With this configuration, only 2 or 3 sample rotations are required to get a good coverage of pole figures with collection times of few minutes (cf. Fig. 17).

Rq.: Neutron imaging techniques may also seem promising for characterizing crystallographic textures [34, 35]. Indeed, diffraction affects the neutron transmission through a polycrystalline material: this is evidenced by the presence of sharp steps at specific wavelengths in the transmission spectrum, the so-called Bragg edges. A Bragg edge is due

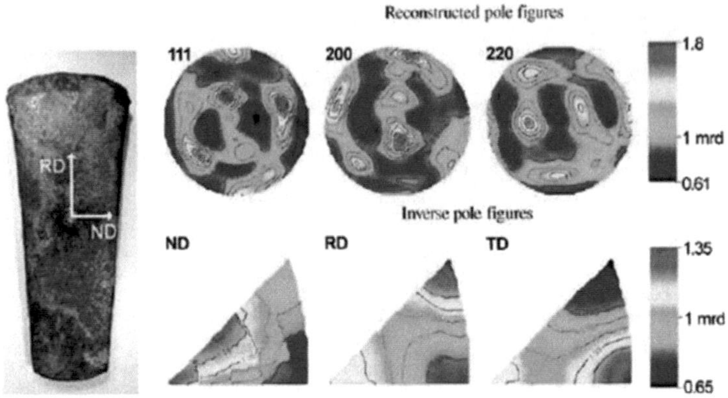

Figure 18. Left: one the 20 prehistoric copper axes which have been analyzed by neutron diffraction (this one has been found in Montecchio, Italy); Right: obtained PFs and inverse PFs evidencing a copper-like texture due to mechanical working along TD followed by thermal recrystallization [36] (*mrd = multiple of random distribution*).

to the impossibility to fulfill the Bragg law for a given hkl reflection if the wavelength λ is greater than $2d_{hkl}$: when $\lambda \leq 2d_{hkl}$, diffraction on {hkl} planes can occur, thus contributing to the total absorption cross section, but when $\lambda \geq 2d_{hkl}$, a sharp increase in the transmitted intensity is generally observed. The height of each step is theoretically proportional to the volume of crystallites having (hkl) planes perpendicular to the incident neutron beam: measuring various hkl Bragg edges for various sample orientations should thus allow to quantitatively analyze the crystallographic texture and its spatial variations. However, only qualitative analysis of texture variations across various materials are reported so far, but improvement of TOF techniques at pulsed sources for neutron imaging may make possible quantitative characterizations in the near future.

5. Examples

5.1 Archaeological metals [36, 37]

Neutron diffraction is an extremely valuable tool for the characterization of materials related to cultural heritage, since the large penetration of neutrons in most materials generally allows to perform measurements on relatively large, massive objects which are mostly precious, without having to cut and take small samples from them, and without specific surface preparation, i.e., non destructively.

As a first example, the manufacturing processes of various prehistoric copper axes (dated the Copper age) have been investigated by means of texture measurements by neutron diffraction [36]. Experiments were performed on entire objects either on D20 at ILL, with a monochromatic wavelength (the analyzed volume being defined by cadmium masks placed on the sample surface), or on GEM at ISIS, using TOF technique (the analyzed volume being defined by cadmium slits on the incoming and diffracted beams). On D20, 6 hours were needed per sample to record 1248 diffraction patterns, corresponding to 1248 sample orientations (ω, χ, ϕ), whereas, on GEM, only 2 sample orientations were sufficient, reducing the counting time to ~ 40 min per sample. A photograph of an ax sample, as well as its corresponding pole figures (and inverse pole figures) are reproduced on Fig. 18. As main results deduced from the qualitative analysis of PFs, it was shown that most axes have been casted, cold worked, with low reduction rates in the range 5%–10%, and subsequently

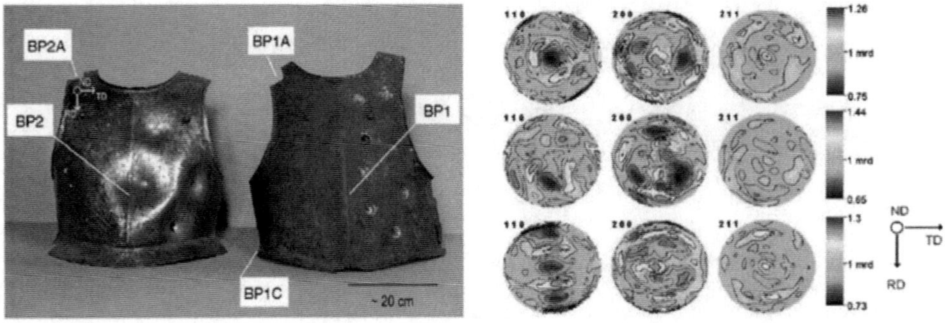

Figure 19. Left: the two harquebusier breastplates with the analysis spots indicated. Right: pole figures recorded on GEM diffractometer from three fragments BP1A (top), BP1C (middle) and BP2A (bottom) [37].

annealed and softened (most are partly recrystallized). Authors deduced that, in the Copper age, mechanical working was not used to harden the metal but rather probably to mask surface casting defects and porosity.

Another quite amusing example deals with the investigation of the manufacturing process of harquebusier breastplates [37]. Textures of two breastplates (BP1 and BP2) from different sources, both initially supposed dated round 1600–1650 AD, were measured on GEM diffractometer (Fig. 19). Diffraction showed that both breastplates were made of ferritic steels, with small quantities of carbides and oxides detected in BP1. Pole figures obtained from BP1 are rather irregular, whereas those obtained from BP2 exhibit typical features of the rolling texture of ferritic steel. The armourers of the 16th and 17th century made their objects by hammering iron blooms into thin metal sheets: the random texture observed in BP1 is thus probably due to random stacking of several hammered iron sheets. However, rolling process was unknown in the 17th century! Moreover, relatively large amounts of Mn (0.31 wt.%) were found from chemical analysis of BP2, but such high Mn contents have never been observed in steels elaborated during the 17th century. Authors thus concluded that BP2 must be a replica, possibly from around 1850–1900 and made for use in the Victorian period.

5.2 Deformation and recrystallization textures in cold wire-drawn copper [38]

Static recrystallization of plastically deformed metallic materials can occur during a high temperature annealing, and lead to the emergence of a new microstructure, generally associated with important evolutions of macroscopic mechanical properties. Its experimental and theoretical study thus still remains an important research area. Two main theories are referred to, which are subjects of debates: the "oriented nucleation" theory is based on a preferential nucleation of grains with specific orientations within the deformed matrix, whereas the "oriented growth" theory considers that nuclei in particular crystallographic orientation relationships with the neighboring grains should grow preferentially.

Due to the statistically representative information it provides, neutron diffraction has been widely applied to follow texture evolution during recrystallization. For instance, deformation and recrystallization textures of cold wire-drawn copper were studied in LLB, especially by taking advantage of a special furnace mounted on the cradle of 6T1 diffractometer to perform in situ diffraction measurements [38].

Copper (purity 99.99%) has been hot rolled and cold wire-drawn at several reduction between 51 and 94%, then annealed at 260°C. Deformation textures exhibit strong a <111> fiber and a weaker <100> fiber (Fig. 20), the texture sharpness increasing with

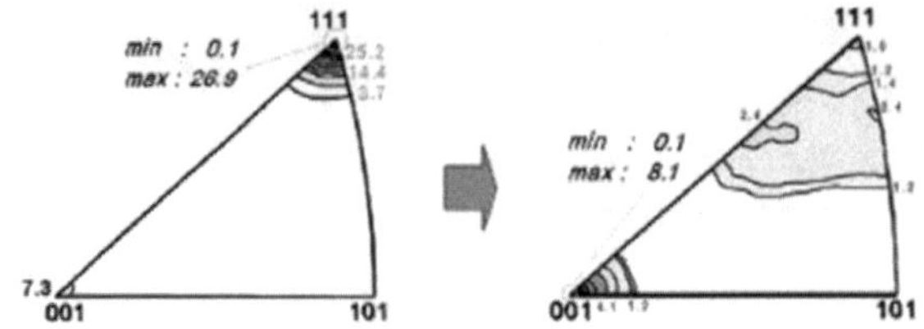

Figure 20. Inverse pole figure (in wire direction) of wire-drawn copper at 94% reduction: as deformed (left) and recrystallized (right) [38]. (*an inverse PF plots a selected sample direction in the CAS, whereas a direct PF plots the* $\mathbf{Q}_{hkl}$ *direction in the SAS*).

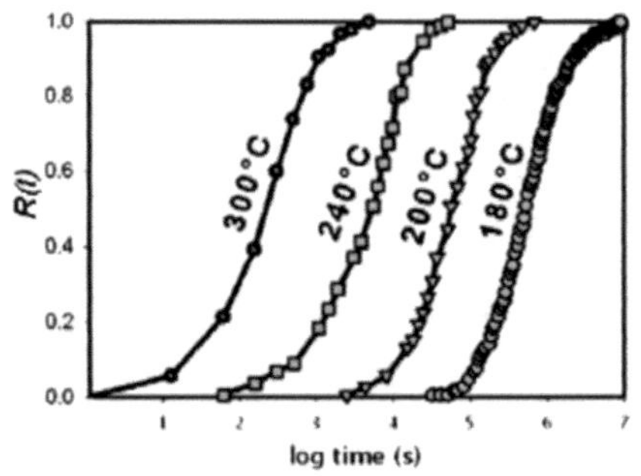

Figure 21. Reaction avancement factor calculated from measurements of the diffracted intensity at the center of 111 PF, in situ at various temperatures [38].

the deformation. After the thermal treatment at 260°C, the recrystallization textures are characterized by a dominant <100> fiber, with a weak <111> fiber which disappears at the highest reduction rates while low volume fraction of <112> and <122> fibers develop (Fig. 20).

The recrystallization process was then followed in situ by recording the evolution with time $I(t)$ of the 111 diffracted intensity at ($\chi = 0°, \phi = 0°$), at constant temperatures. The reaction advancement factor $R(t)$ of the recrystallization process was thus calculated by:

$$R(t) = \frac{I(0) - I(t)}{I(0) - I(\infty)} \tag{14}$$

The evolution of $R(t)$ at various temperature for 94% reduction is plotted in Fig. 21. Static recrystallization is a thermally activated process, and $R(t)$ variations at different temperatures allowed to compute the activation energy of the recrystallization process as a function of the strain level. An exponential decay of the activation energy with increasing strain could thus be deduced, reaching values close to 45 kJ/mol at reductions $\geq 90\%$.

5.3 Deformation of beryllium [39]

Plastic deformation of hexagonal close packed (hcp) metals such as Be, Mg or Zr is characterized by the possible activity of different deformation modes, such as dislocation glide on various slip systems (dislocations may move on the basal {00.1}, prismatic {10.0} or pyramidal {10.1} planes) or deformation twinning. Crystallographic texture, temperature and strain rate may strongly affect the relative contributions of these deformation modes. In order to better understand the deformation mechanisms of Be as a function of texture and

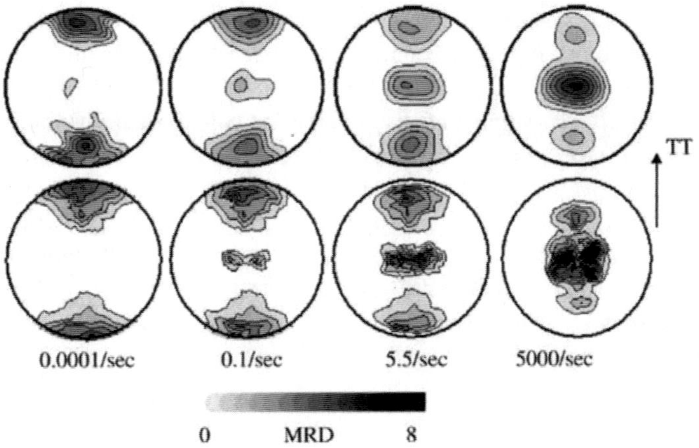

Figure 22. Measured (top) and simulated (bottom) 00.1 pole figures for the deformation textures of rolled Be at various strain rates (up to 22% compressive strain, in the direction normal to the page). TT: Through Thickness direction [39].

strain rate, Brown et al. [39] used a coupled neutron diffraction and polycrystalline modelling approach.

Be samples with two distinct starting textures (either nearly random obtained by hot pressing or strongly textured obtained by rolling) were deformed in uniaxial compression at various strain rates (covering a wide range from 0.0001/s to 5/s). Initial and deformation textures were measured on HIPPO diffractometer (LANSCE). A multi-scale constitutive model, which accounts for slip and twinning in each grain, as well as twin reorientation, was applied to simulate stress-strain curves and deformation textures as a function of the strain rate, and the latter were confronted to the experimental results. For instance, Fig. 22 compares the measured and calculated {00.1} deformation pole figures of the rolled (initially textured) Be after in-plane compression. Authors managed to satisfactorily reproduce the observed flow curves as well as the texture evolutions for all mechanical tests with a unique set of parameters. This experiment – modeling confrontation allowed them to compare the rate sensitivities of the different deformation modes. This should lead to a better understanding of the roles played by plastic anisotropy, and by the competitions between various slip systems and twinning in the mechanical properties of hcp metals.

6. Conclusions

Crystallographic texture is an important microstructural parameter which directly determines the anisotropy degree of most physical properties of a polycrystalline material at the macro scale. Its characterization is thus of fundamental and applied importance, and should ideally be performed prior to any physical property measurement or modeling. In material science, understanding the physical properties – microstructure relationship in materials remains a very active research field, especially in mechanics: theoretical approaches still have to be developed and experimentally validated to solve the complex scale transition problem of passing from a single crystal anisotropic property to a polycrystal property, by taking into account the full description of the microstructure, i.e., not only the crystallographic texture, but also the grain morphologies (size and shape distributions), the spatial arrangement of the crystallites (e.g., with correlation functions), the nature of grain boundaries, etc and of course the nature and distributions of the constitutive phases and inclusions, porosities, etc.. etc..

Neutron diffraction is a tool of choice for characterizing crystallographic textures. Its main advantages over other existing techniques, and especially over the X-ray diffraction techniques, are essentially due to the low neutron absorption by most elements. Hence, sample preparation and data treatment are generally easy, bulk measurements are feasible, and complex sample environments (e.g., for in situ studies) can be used as well. Relatively low available neutron fluxes (if compared with X-ray sources) are compensated by the utilization of large neutron beams (> mm^2): the obtained information is thus representative of a large number of grains, leading to a better accuracy of the statistical description of texture. Neutron diffraction is also especially well suited to the characterization of materials with particular microstructural features, such as coarse grains, multiple phases, minority phases.

References

[1] U.F. Kocks, C.N. Tomé, H.R. Wenk, *Texture and Anisotropy - Preferred Orientations in Polycrystals and their Effect on Materials Properties* (Cambridge University Press, 1998)
[2] J.F. Nye, *Physical Properties of Crystals; their representation by tensors and matrices* (Oxford University Press, 1957)
[3] D. Sander, Z. Tian, J. Kirschner, Sensors **8**, 4466 (2008)
[4] S. Matthies, G.W. Vinel, K. Helming, *Standard Distributions in Texture Analysis - Maps for the Case of Cubic-Orthorhombic Symmetry*, Vol. 1 (Berlin: Akademie-Verlag, 1987)
[5] U.F. Kocks, in *Eighth Int. Conf. on Textures of Materials*, edited by J. Kallend, G. Gottstein (1988), pp. 31–36
[6] H.R. Wenk, U.F. Kocks, Metall. Trans. **18A**, 1083 (1987)
[7] H.J. Bunge, *Texture Analysis in Materials Science – Mathematical Methods* (London: Butterworths, 1982)
[8] R.J. Roe, J. Appl. Phys. **36**, 2024 (1965)
[9] O. Rodrigues, J. Mathématiques Pures et Appliquées **5**, 380 (1840)
[10] A. Morawiec, J. Pospiech, Textures Microstruct. **10**, 211 (1989)
[11] H.J. Bunge, Z. Metallk. **56**, 872 (1965)
[12] H.J. Bunge, C. Esling, J. Phys. Lett. **40**, 627 (1979)
[13] C. Esling, H.J. Bunge, J. Muller, J. Phys. Lett. **41**, 542 (1980)
[14] H.J. Bunge, C. Esling, J. Muller, J. Appl. Crystallogr. **13**, 544 (1980)
[15] H.J. Bunge, C. Esling, J. Muller, Acta Crystallogr. **A37**, 889 (1981)
[16] H.J. Bunge, C. Esling, Acta Crystallogr. **A41**, 59 (1985)
[17] D. Ruer, R. Baro, Adv. X-Ray Analysis **20**, 187 (1977)
[18] S. Matthies, G.W. Vinel, Phys. Stat. Sol. (b) **112**, K111 (1982)
[19] K. Pawlik, J. Pospiech, K. Lücke, Textures Microstruct. **14–18**, 25 (1991)
[20] B.D. Cullity, *Elements of X-Ray Diffraction (2nd Edition)* (Addison-Wesley, Reading, 1978)
[21] C. Esling, H.J. Bunge, Techniques de l'Ingénieur **TIB476DUO**(M3040) (2015)
[22] G. Ischia, H.R. Wenk, L. Lutterotti, F. Berberich, J. Appl. Cryst. **38**, 377 (2005)
[23] D. Juul Jensen, H.F. Poulsen, Mater. Charact. **72**, 1 (2012)
[24] J.S. Chung, G.E. Ice, J. Appl. Phys. **86**, 5249 (1999)
[25] H. Rauch, W. Waschkowski, in *Landolt-Boernstein* (Springer, Berlin, 2000), Vol. 16A, Chap. 6
[26] H.R. Wenk, S. Matthies, L. Lutterotti, in *Textures of Materials ICOTOM-10*, edited by H.J. Bunge (Switzerland: Trans. Tech. Pubs., 1994), Vol. 157–162, pp. 473–479

[27] L. Lutterotti, S. Matthies, H.R. Wenk, A.J. Schultz, J.W. Richardson Jr., J. Appl. Phys. **81**, 594 (1997)

[28] R.B. Von Dreele, J. Appl. Cryst. **30**, 517 (1997)

[29] `www-llb.cea.fr/`

[30] H.G. Brokmeier, W.M. Gan, C. Randau, M. Völler, J. Rebelo-Kornmeier, M. Hofmann, Nucl. Instrum. Method. Phys. Res. **642**, 87 (2011)

[31] `www.ill.eu/`

[32] S.C. Vogel, C. Hartig, L. Lutterotti, R.B. Von Dreele, H.R. Wenk, D.J. Williams, Adv. X-Ray Analysis **47**, 431 (2004)

[33] W. Kockelmann, L.C. Chapon, P.G. Radaelli, Physica B **385–386**, 639 (2006)

[34] J.R. Santisteban, L. Edwards, V. Stelmukh, Physica B **385–386**, 636 (2006)

[35] J.R. Santisteban, M.A. Vicente-Alvarez, P. Vizcaino, A.D. Banchik, S.C. Vogel, A.S. Tremsin, J.V. Vallerga, J.B. McPhate, E. Lehmann, W. Kockelmann, J. Nucl. Mater. **425**, 218 (2012)

[36] G. Artioli, Appl. Phys. A **89**, 899 (2007)

[37] S. Leever, D. Visser, W. Kockelmann, J. Dik, Physica B **385–386**, 542 (2006)

[38] P. Gerber, S. Jakani, M.H. Mathon, T. Baudin, Mat. Sci. Forum **495–497**, 919 (2005)

[39] D.W. Brown, I.J. Beyerlein, T.A. Sisneros, B. Clausen, C.N. Tomé, Int. J. Plast. **29**, 120 (2012)

EPJ Web of Conferences **155**, 00006 (2017) DOI: 10.1051/epjconf/201715500006
JDN 22

Diffuse scattering

Isabelle Mirebeau[a]

Laboratoire Léon Brillouin, CEA-CNRS, CE-Saclay, 91191 Gif-sur-Yvette, France

Abstract. Diffuse neutron scattering covers a wide range of phenomena related to short range nuclear and magnetic orderings. Although it is "a priori" simple to measure, the underlying physics is often quite complex. Extracting useful and reliable information requires careful corrections and calibrations, and appropriate models of analysis, specifics for each physical case. This paper yields a partial and subjective glance on this specific subject, showing studies about chemical orderings in binary alloys and magnetic correlations in frustrated "spin ices" as examples.

1. Introduction

In reciprocal space, the term "diffuse scattering" covers scattering processes which extend on a large part of the Brillouin zone. It therefore contrasts with "Bragg scattering" which is limited to the vectors of the reciprocal space, and with "small angle scattering" which covers a small portion of the reciprocal space surrounding the zone center (typically up to 0.1 Å^{-1}). Neutron scientists have usually reserved the term of "diffuse scattering" to a diffraction process, measured in the so called "two axis" geometry, without performing an energy analysis of the scattered neutron beam. For fluctuating phenomena, this means that one basically measures the instantaneous correlations, taking a "snapshot" of the system. In many cases however, the process of interest is of elastic nature, as it involves static configurations. Recent measurements also select the elastic contribution by performing an energy analysis of the scattered beam using time of flight techniques for instance.

In real space, diffuse scattering allows one to investigate a large variety of states of imperfect matter, where defects range at a local scale, in opposition to the long range periodic order and to the mesoscopic order studied through Bragg and small angle scattering processes respectively. Actually any real sample is imperfect when one studies it sufficiently closely, and the experiment looks quite simple: one just measures everything between the Bragg peaks! Therefore, for a long time, diffuse scattering measurements have been considered in a rather despising way as "scraping the background". The difficulty arises when one wants to extract reliable information from this diffuse "background", coming from the phenomenon of interest. This requires extracting the relevant signal from all other sources of environmental and physical background contributions. This procedure may require the calibration of the measured signal in absolute scale, and a comparison with an "*a priori*" model to refine the relevant parameters. Such refinements may be complex, since there is no general procedure such as those developed to analyze long range periodic order, so that each physical case is specific and yields its own method of analysis.

Which states of matter can be studied through diffuse scattering? Actually there are no limitations in the states of disorder or local order which can be investigated in solid

[a] e-mail: isabelle.mirebeau@cea.fr

state physics. The great advantage of the neutron probe is the possibility to investigate the bulk state, and to study nuclear and magnetic correlations separately. As for nuclear scattering, one can quote studies of local occupation disorder in solid solutions and alloys, initiated by the pioneering work of Cowley and Warren [1], static local displacements initially modelled by Borie and Sparks [2], orientation disorder, local defects in quasi-crystals or aperiodic materials [3] among others. In all these studies, an underlying periodic or quasi-periodic lattice is postulated. Amorphous and liquid states also yield diffuse scattering, but specific tools of analysis have been developed in such cases which will not be mentioned here. As for the magnetic scattering, one can quote studies of magnetic correlations above a critical transition, the determination of local moments and local perturbations in ferromagnetic alloys [4], investigations of 2D and 1D systems [5], Bose condensates with dimer ordering [6], magnetic fluctuations in unconventional superconductors [7], and the large variety of entropic disordered ground states observed in frustrated systems, such as spin liquids, spin ices, or spin glasses [8], among others.

In the following, this chapter starts to provide a few guidelines about the necessary corrections, which could be helpful for who discovers the technique, and briefly mentions the general descriptions developed as a first approach, depending on the reference state (fully ordered or fully random). An example of the progressive evolution from order to disorder is shown by the quantum critical transition from canted ferrimagnetism to spin-glass states in molybdate pyrochlores. Then one focuses on a few examples where neutron diffuse scattering was modeled and yielded useful information. These examples deal with binary alloys with site disorder, and with the exotic local magnetic order shown by the "spin ices".

2. How to measure diffuse scattering? Experimental corrections

Measurements of the diffuse scattering are usually done on a two axis spectrometer. For powder, one uses a powder diffractometer equipped with a multidetector, whereas for single crystals several geometries can be used. The most common one involves multi-counters in the scattering plane such as D7 at the ILL [9], allowing one to scan a given fraction of the desired scattering plane by rotating an oriented crystal along a vertical axis (the so-called "snail curves"). In some cases, polarized neutrons are used to separate nuclear from magnetic scattering contributions. In ferromagnetic alloys, unpolarized neutron measurements combining different orientations of the saturating magnetic field are an interesting alternative.

The procedure to determine the diffuse scattering contributions arising from magnetic and nuclear short range order in a ferromagnetic alloy is given here as an example of the typical corrections which should be performed. The sample is measured at low temperature, typically 2 to 5 K, to eliminate the contribution of phonons, and for two geometries of the magnetic field **H**: in zero field corresponding to random orientations of the ferromagnetic domains; under a saturating field $\mathbf{H}_{sat}$ perpendicular to the scattering plane, to obtain a single domain oriented perpendicular to all **Q** vectors.

The intensities of the empty sample holder (E), vanadium sample (V), and fully absorbing cadmium sample (Cd) are also measured with the same experimental set-up. To minimize the corrections, Vanadium and Cadmium samples occupy the same volume as the studied sample (assumed here to be a cylinder with radius R and length L, with $L \gg R$). The following formula accounts for: 1) the correction from the environmental background due to the neutron beam after traversing the sample (E-Cd) and from the background outside the sample (Cd); 2) the detector efficiency; 3) the absorption correction [10]; 4) the multiple scattering corrections [11]; 5) an absolute calibration, taking the incoherent scattering of vanadium as a reference.

Absorption and multiple scattering corrections are tabulated for a cylindrical sample and for standard geometries, knowing the absorption coefficient $\mu = n\sigma_T$ (where n_S is the sample density and σ_T the *total* sample cross section estimated for the incident neutron wavelength of the experiment, from the cross sections of the individual constituents) [12]. In the case of complicated shapes, a Monte-Carlo correction has to be used to account for the absorption and multiple scattering of all trajectories inside the sample, given the incident and exit directions. Corrections taking inelastic scattering into account through Monte Carlo codes have been also developed [11].

The corrected intensity scattered by the sample N_S^{cor} is given by:

$$N_S^{cor} = N_S - N_{Cd} - A_S(0)[N_E - N_{Cd}]$$

where N_S, N_E and N_{Cd} are the intensities recorded from the sample, the empty sample holder and a Cd sample of the same volume respectively, for the same number of incident neutrons. A(0) is the transmission coefficient.

The neutron scattering cross section is given in absolute units by:

$$\frac{d\sigma}{d\Omega}(\mathbf{Q}) = \frac{\sigma_{inc}}{4\pi}\frac{n_V}{n_S}\frac{A_V(\theta)}{A_S(\theta)}\frac{(1-\delta'_S)}{(1-\delta'_V)}\frac{N_S^{cor}(\theta)}{N_V^{cor}(\theta)}$$

where $\sigma_{\text{inc}} = 5.07$ barn is the incoherent cross section of vanadium; n_S and n_V are the sample and vanadium number of atoms, as deduced from the mass of the measured samples; A(θ) is the absorption coefficient and 2θ the scattering angle; δ'_S and δ'_V are the multiple scattering coefficients of the sample and vanadium.

From the total cross section, the magnetic and nuclear diffuse cross sections are given by:

$$\frac{d\sigma}{d\Omega}(\mathbf{Q}) = \frac{d\sigma_N}{d\Omega}(\mathbf{Q}) + \frac{2}{3}\frac{d\sigma_M}{d\Omega}(\mathbf{Q}) \quad for\ \mathrm{H} = 0$$

$$\frac{d\sigma}{d\Omega}(\mathbf{Q}) = \frac{d\sigma_N}{d\Omega}(\mathbf{Q}) + \frac{d\sigma_M}{d\Omega}(\mathbf{Q}) \quad for\ \mathbf{H} \perp \mathbf{Q}$$

some other configurations can be used: i) measurements in a field along the scattering vector **Q** (**H // Q**); ii) polarized neutron measurements, with $\mathbf{P}_Z$ either parallel or antiparallel to the saturating field (**H** $\perp$ **Q**). They yield the following equations:

$$\frac{d\sigma}{d\Omega}(\mathbf{Q}) = \frac{d\sigma_N}{d\Omega}(\mathbf{Q}) \quad for\ \mathbf{H}//\mathbf{Q}$$

$$\frac{d\sigma}{d\Omega}(\mathbf{Q}) = \frac{d\sigma_N}{d\Omega}(\mathbf{Q}) + \frac{d\sigma_M}{d\Omega}(\mathbf{Q}) \pm \frac{d\sigma_{NM}}{d\Omega}(\mathbf{Q}).$$

These equations may be combined to extract the differential cross sections $\frac{d\sigma_N}{d\Omega}(\mathbf{Q})$, $\frac{d\sigma_M}{d\Omega}(\mathbf{Q})$, $\frac{d\sigma_{NM}}{d\Omega}(\mathbf{Q})$ corresponding to the nuclear, magnetic and interference terms respectively.

Using this procedure, one can calibrate the diffuse intensity without using the Bragg scattering as a reference. This is important since there might be no Bragg scattering from the sample if one uses a large incident wavelength to measure the diffuse scattering with a good resolution. Moreover, diffuse scattering measurements are usually performed on large single crystals in “two-axis” geometry, so the Bragg intensities can be strongly affected by extinction effects or by imperfect sample orientation. This method was used to study the local chemical and magnetic orders in solid solutions of Fe, Ni or Co based ferromagnetic alloys, among others.

In the case of dilute Fe-Cr alloys (see paragraph 4 below), where the nuclear signal from local Cr order is very small due to the small contrast between Cr and Fe atoms, the same $Fe_{0.95}Cr_{0.05}$ sample was measured twice, with different background conditions [13].

In the first set-up the intensity coming from the sample was well below that of the environmental background (mostly the air scattered by the incident neutron beam) and the sample signal could not be detected without performing all corrections. By using a vacuum chamber compatible with a cryostat and an electromagnet, the second set-up decreased the environmental background by about a factor 20, allowing a direct visualization of the sample signal, a better precision of the neutron cross section yielding the determination of short range order parameters over further neighbor distances, and measurements at lower Cr contents. Although the statistical precision was significantly improved in the second case, suitable corrections and calibrations yielded the same value of the average cross section in absolute scale for both set-ups, which compares well with the calculated sum of the incoherent and Laue scattering cross sections. This allowed the nearest neighbor short range order parameters to be determined in any case.

3. How to describe diffuse scattering? A few guidelines

Usually the diffuse scattering signal is described either as a deviation from the Long Range Ordered state (LRO yielding Bragg scattering) or as a deviation from the fully disordered or random state (yielding incoherent Laue scattering or uncorrelated paramagnetic scattering). In other words, the Q dependence of the diffuse intensity I(Q) is simulated either by a broadening of the Bragg peaks, or by modulations of the flat incoherent "background" between them. The first method which takes LRO as a reference yields the determination of correlations lengths, typical domain size, critical behaviors [14], long ranged distortions [15], in some cases quantum fluctuations or spin waves constants [5, 16]. It is well adapted to typical length scales of a few tens of unit cells. The second method, taking the random state as a reference, yields the determination of short range order nuclear parameters [1, 13, 17–20], short range ordering of vacancies [21], local magnetic perturbations [22], local lattice distortions [2, 21], or pair distribution functions [23]. It covers atomic or spin correlations mostly limited to near neighbor distances. Examples of such cases are shown in paragraph 4 for $Fe_{1-x}Cr_x$ polycrystalline alloys and a FePd single crystal.

The choice of the appropriate description and reference state obviously depends on the context. As an example, Fig. 1 shows the evolution of the magnetic scattering from LRO to spin glass states across a quantum critical transition. This quantum transition which occurs in molybdate pyrochlores $(Tb_{1-x}La_x)_2Mo_2O_7$ can be encompassed either by *La* substitution which expands the lattice (Fig. 1 left) or by applied pressure which compresses the lattice (Fig. 1 right) [24]. Expanding the lattice yields a transition from spin glass to LRO, and the reverse occurs by lattice compression, due to the combined effects of a Mott transition and geometrical frustration [25]. At ambient pressure, in the spin glass sample ($x = 0$), the large sinusoidal modulations can be fitted by local magnetic parameters [26], yielding correlations over a few interatomic distances. Such correlations are either ferromagnetic (F) or antiferromagnetic (AF) depending on the magnetic pair concerned. At intermediate concentrations ($x = 0.1$), an intermediate mesoscopic order shown by Lorentzian line shapes develops over several unit cells at the positions of the Bragg peaks of the LRO, present in the $x = 0.2$ sample. This canted ferromagnetic LRO can be determined by refining the pattern of the $x = 0.2$ sample with the Fullprof suite [27], yielding the moment values and orientations in the ordered magnetic unit cell. The evolution of the peak line shapes shows that the two processes, chemical pressure versus applied pressure, are not fully equivalent. When applying pressure, the LRO is destroyed in an inhomogeneous way, likely due to pressure inhomogeneity, whereas a homogenous transition occurs under *La* substitution. The pattern of the $x = 0.2$ sample at 1.9 GPa shows the coexistence of Bragg peaks and sinusoidal modulations, respectively associated with LRO and near neighbor correlations. In contrast

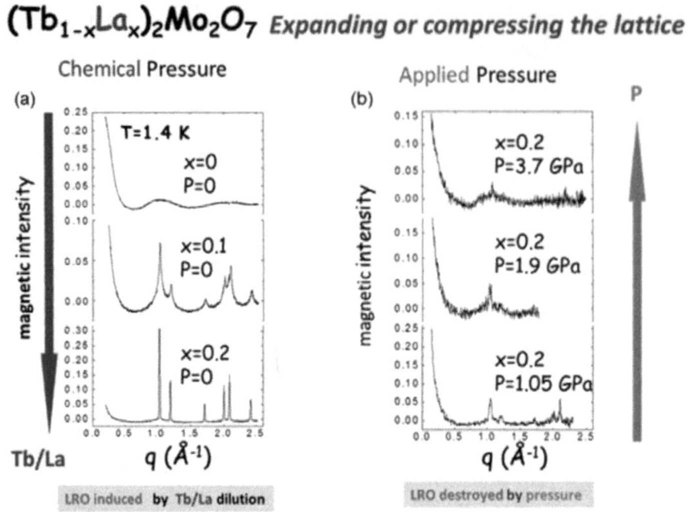

Figure 1. Evolution of the diffuse scattering through a quantum transition from spin glass to long range order in $(Tb_{1-x}La_x)_2Mo_2O_7$ pyrochlore alloys. The quantum transition can be encompassed either a) by chemical pressure (*La* substitution) or b) applied pressure. Note that the processes are different in the two cases, as shown by the intermediate magnetic phases at the middle part of the figures. Under chemical pressure ($x = 0.1, P = 0$), a mesoscopic phase appears with finite correlation length, as shown by the Lorentzian shape of the magnetic peaks. Under applied pressure ($x = 0.2, P = 1.9$ GPa), the intermediate state consists of short-range correlated regions (at the scale of a few interatomic distances) coexisting with a long-range magnetic ordering that involves a small ordered moment; from Ref. [24].

Lorentzian line shapes are observed in the $x = 0.1$ sample at ambient pressure, which can be related to the onset of homogeneous ordering over an intermediate length scale.

More recently new techniques have been developed to describe the diffuse scattering. For nuclear diffuse scattering, the determination of the pair distribution function (pdf) inspired by the studies of amorphous and liquid alloys, yields atomic pair correlations, interatomic distances, Debye-Waller coefficients and occupation numbers [26, 28]. It starts to be extended to the case of magnetic alloys [29]. Other methods based on Reverse Monte Carlo techniques [30] also allow one to obtain local magnetic correlations, assuming an energetic model and adjusting its parameters to obtain the best agreement with experiment. This is also the case of the inverse cluster variation method [31].

In some geometrically frustrated systems, such as spin ices, spin liquids or quantum spin ices, an original description of the short range ordered ground state and associated diffuse scattering has been very successful [8, 32, 33]. It is based on the postulate of conservation laws (such as the ice laws), yielding emergent electrodynamics and gauge fields. Postulating such laws is crucial, as it yields a description of a magnetic state different both from the paramagnetic state and from a long range ordered ground state. By giving a local conservation rule, one can define a fictitious field, obeying the general laws of electrodynamics. For classical Ising spins obeying "ice rules", this conservation law appear as a Gauss's law, and the associated electric field is directly linked with the classical spin. The dipolar character of the electric field in real space results in singularities of the magnetic structure factor in Fourier space. These highly anisotropic features show up as "bow-tie" features and "pinch-points", which can be directly observed in the Q-dependence of the neutron diffuse scattering. Flipping an individual spin violates the ice rules. These thermally activated excitations induce "magnetic monopoles", either source or sink of magnetic flux, which appear as a broadening of the pinch points in the diffuse maps. For quantum spins, transverse energy terms (with respect to the Ising character), result in an effective field which has both electric and magnetic degree of freedom. The predicted excitations behave as particles of light. Observing these "emergent photons" in quantum spin ices constitutes a challenge of modern condensed matter physics. A brief insight on this fascinating subject, with some examples of the neutron diffuse scattering maps is given in paragraph 5.

4. Local chemical ordering in binary solid solutions

4.1 $Fe_{1-x}Cr_x$ alloys

$Fe_{1-x}Cr_x$ alloys are a unique case in nature showing an inversion of the short range atomic order within a solid solution. Namely the Cr atoms in a Fe matrix repel each other at low concentration, whereas they attract each other at high concentration. This feature seems to be at the origin of the peculiar mechanical properties shown by $Fe_{1-x}Cr_x$ alloys in the solid solution, which have important mechanical applications. There is presently a considerable effort in the search for materials highly resistant to radiation damages, which will be used both in future hybrid-type reactors and in future controlled fusion reactors. $Fe_{1-x}Cr_x$ alloys with base centered cubic (bcc) structure are the reference model to understand the behavior of ferritic FeCr based steels, considered as leading candidates in most future nuclear energy options. As a striking fact, the response of $Fe_{1-x}Cr_x$ alloys over irradiation is highly non monotonic, with a pronounced change of behavior around the critical concentration $x_c \sim 0.10$, where the short range order inversion takes place.

Understanding this behavior requires a precise knowledge of the pair interactions between the atoms which govern the distribution of the atoms over the lattice sites. In a solid solution, the atomic distribution is never fully random, and peculiar local configurations are privileged in a statistical way, controlled by the electronic configuration. In transition metal alloys, it has been known for decades that the electronic configuration energy can be written in terms of pair interactions [34].

Neutron diffuse scattering is the best way to determine the short range magnetic order in a solid solution. This is especially true for transition metal alloys of the first series, where the electronic contrast (difference of electronic numbers) that governs the amplitude of the X ray diffuse scattering intensity is quite small. Following a theoretical prediction based on the electronic band structure, neutron measurements were the first probe to observe the surprising inversion in $Fe_{1-x}Cr_x$. They were later confirmed by numerous experimental studies based on resistivity measurements, Mössbauer effect, thermoelectric power, enthalpy of formation, bulk modulus, irradiation by heavy ions, etc.. The initial band structure calculations were also followed by improved *ab initio* theories as well as extended Monte-Carlo simulations, confirming the validity of the pair potential approach.

How are short range order effects and short range order inversion shown up by neutron diffuse scattering? The visualisation of short range order is quite simple, since the nuclear cross section is the Fourier transform of the short range order Cawley-Warren parameters, which describe the statistical deviation of a local configuration with respect to that of a random distribution [1].

In a binary alloy $A_{1-x}B_x$, taking an A type atom at the origin, the atomic SRO parameter α_i is related to the conditional probability p_{AB} of having a B atom at a distance R_i:

$$\alpha_i = 1 - \frac{p_{AB}}{x} = 1 - \frac{p_{BA}}{1-x}.$$

A positive α_i value describes a tendency towards segregation ($p_{AB} < x$) over the ith neighbour shell surrounding the A atom, whereas a negative value corresponds to a short range order

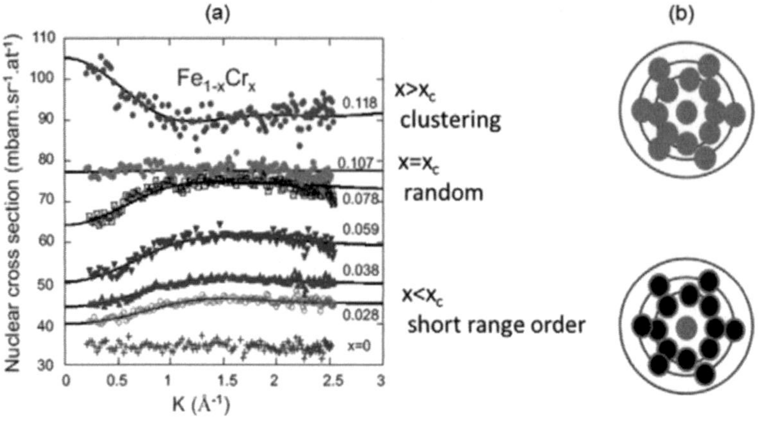

Figure 2. a) Neutron diffuse scattering in $Fe_{1-x}Cr_x$ alloys for different Cr contents x. A transition occurs at the critical concentration $x_c = 0.10$ At low concentration ($x<x_c$) the alloys tends to short range ordering (the neutron cross section decreases at low Q values). For $x=x_c$, the atomic distribution is random (the neutron cross section is flat). For $x>x_c$, the system tends to segregate (the neutron cross section increases at low Q values); b) schematic representation of the atomic distribution around a Cr atom (in red). From Ref. [13].

of hetero-coordination type ($p_{AB} > x$). In a single crystal, the neutron cross section is expressed by:

$$\frac{d\sigma}{d\Omega} = x(1-x)(b_A - b_B)^2 \sum \alpha(\boldsymbol{R_n}).\cos(\boldsymbol{Q}.\boldsymbol{R_n}).$$

Where b_A and b_B are the coherent scattering lengths associated with A and B species, and the sum extend over all pairs separated by a vector $\mathbf{R_n}$. One notices that in a single crystal, the intensity is periodic in reciprocal space. A tendency to segregation yields diffuse maxima located around Q = 0 and around the structural Bragg positions. A tendency towards "short range order" yields diffuse maxima located away from the structural peaks, usually around the peaks of an ordered superstructure. In a powder sample, the orientation average of the cosine term induces a loss of periodicity in the cross section, which is expressed by:

$$\frac{d\sigma}{d\Omega} = x(1-x)(b_A - b_B)^2 \left[1 + \sum_{i\neq 0}^{n} N_i \alpha_i \frac{\sin Q.R_i}{Q.R_i}\right].$$

Since the amplitude of the SRO parameters decreases with increasing R_i, the behaviour of the nuclear cross section at small Q-values is dominated by the sign of the first parameters. Depending on their sign, the modulations of the cross section suggest either a tendency towards a superstructure (onset of new Bragg peaks) or towards segregation (small angle scattering). Figure 2 shows the nuclear cross sections for a series of $Fe_{1-x}Cr_x$ alloys. At low concentrations, the nuclear cross section decreases at low Q values, whereas it increases at high concentrations. At the critical concentration $x_c = 0.01$, the cross section is Q-independent. This random distribution locates the short range order inversion.

Figure 3 (left) shows the variation of the first atomic SRO parameter α_{12} averaged over the first two neighbour shells. This average was necessary since the first two shells are close to each other in a bcc structure. At low Cr content, α_{12} is close to the limit value $-x/(1-x)$ which corresponds two maximum repulsion between Cr- Cr pairs, and it changes sign at the critical concentration. The experimental variation is compared with calculations from Monte Carlo simulations [35, 36]. In binary alloys, the configuration energy can be written in terms

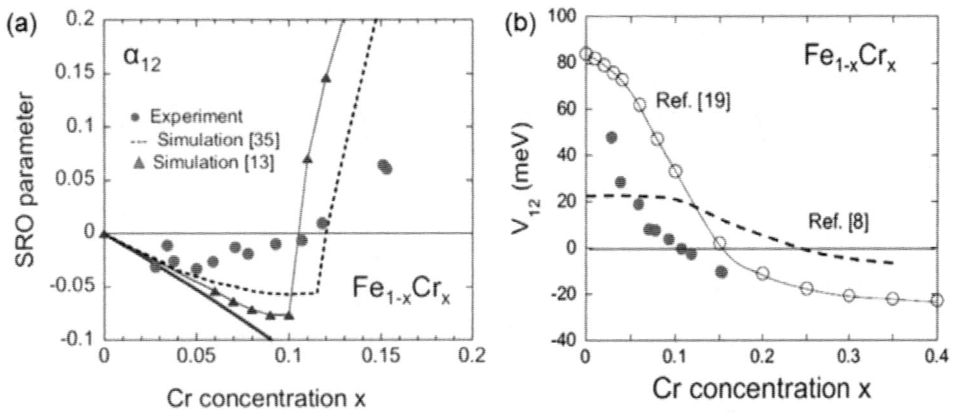

Figure 3. $Fe_{1-x}Cr_x$ alloys: a) variation of the first atomic short range order parameter, averaged over the first two neighbour distances of the base centred cubic lattice with Cr concentration x, from Ref. [13]. The experimental data are compared with the Monte-Carlo simulations of ref. [35, 36]; The solid blue line is the curve $-x/(1-x)$ of maximal repulsion between Cr-Cr pairs; b) variation of the first atomic pair potential V_{12} with Cr content, in comparison with ab initio calculations [38, 39].

of pair potentials and occupation parameters, via an Ising-like Hamiltonian. The minimisation of the free energy in mean field approximation [37] allows one to determine the effective pair potentials which govern the atomic interactions. In $Fe_{1-x}Cr_x$ alloys, the first pair potential V_{12} deduced from the SRO parameters also changes sign at x_c. In Fig. 3 (right), the experimental pair potential is compared with *ab initio* band structure calculations: the predictive behaviour of Ref. [38], and a more recent model [39].

4.2 FePd single crystal

In neutron diffuse scattering studies, the availability of a single crystal opens a route to a much more precise determination of the local order parameters and more detailed comparison with theory. Several extensive studies have been performed in the past, such as the investigation of ternary FeNiCr alloys [17], local order in CoPt [18] or FeV alloys [19], ordered vacancies in NiC [21] among others. A particular interesting case is the investigation of the diffuse scattering for a stoechiometric compound above the transition towards a LRO state. In such case, the analysis of the diffuse scattering allows one to predict the phase diagram, the ordering temperature and the nature of the LRO ground state. This is shown here in the case of cubic FePd [20]. Below the order-disorder temperature ($T_O = 920$ K), FePd alloys around the 50/50 stoechiometry order in the $L1_0$ structure, made of alternating pure Fe and pure Pd (001) planes. This strong anisotropy is accompanied with a magnetic anisotropy and tetragonal distortion. This phase transformation makes FePd alloys interesting as shape memory materials with potential applications as coupling devices.

Diffuse neutron scattering measurements were carried out in two reciprocal planes (100) and (110) at 1020 K, namely above T_O and in the paramagnetic state (well above the Curie temperature $T_C = 760$ K), using time of flight energy analysis to extract the elastic signal from the dominant phonon contribution. The diffuse intensity, shown Fig. 4 is globally periodic in reciprocal space, and mainly located at the 100 and equivalent points, which correspond to the superstructure peaks of the $L1_0$ structure. A small asymmetry of the diffuse spots is due to local static displacements. The contribution from local *magnetic* order can be neglected, besides the ferromagnetic correlations between fluctuating Fe moments around the zone center, which have been discarded from the analysis. The SRO parameters (Fig. 4

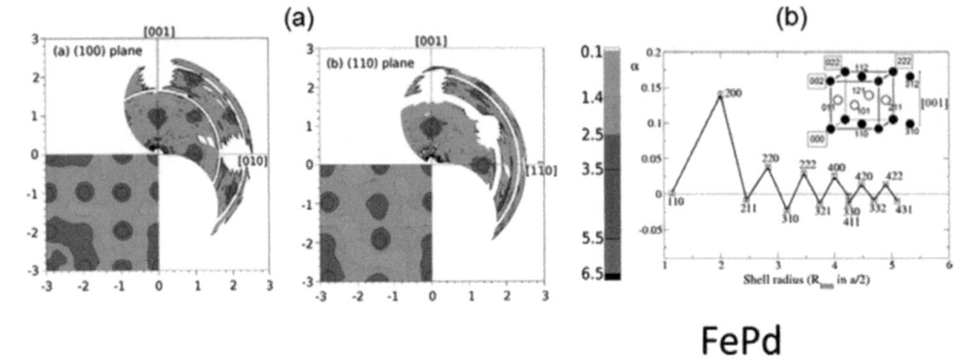

Figure 4. FePd single crystal: a) corrected diffuse experimental intensities in the (100) plane and (110) planes, and reconstructed intensities (down left for each figure) using SRO and lattice displacement parameters; b) variation of the SRO Cawley-Warren parameters versus the neighbour distances. The LRO structure of $L1_0$ type, with alternating pure Fe and pure Pd (001) planes, is shown in inset. From Ref. [20].

right) and lattice displacement parameters deduced from least square fit of the calibrated intensities extend to about 14 and 6 shells respectively. The calculation of the pair potentials from the SRO parameters was carried out using the cluster variation method (CVM) [31] which includes a description of the energy for a large set of clusters, going beyond the Bragg-Williams approximation used in the mean field model of Clapp and Moss [37]. The stability of the $L1_0$ phase is mainly due to the dominant second neighbour pair interactions, and the oscillating behaviour of the interactions between further neighbours.

5. Local magnetic ordering and coulomb phases in spin ices and quantum spin ices

Diffuse scattering has been one of the main tools to study spin ice behaviour in rare earth pyrochlores, starting from the pioneering experiments of Harris and Bramwell in $Ho_2Ti_2O_7$ [40]. In a spin ice, the magnetic moments are constrained by local Ising anisotropy to remain along the local 111 axes of the pyrochlore lattice, made of corner sharing tetrahedra. The minimization of the ferromagnetic exchange interaction within a given tetrahedron leads to the so-called "two in-two out" configuration (namely two spin pointing towards the center of the tetrahedron and two spin pointing outwards). This configuration can be mapped to the configuration which governs the hydrogen positions of water molecules in cubic ice, involving "two short and two long" hydrogen bond lengths in a given tetrahedron. Such configuration has six equivalent realizations. In the pyrochlore lattice of weakly connected tetrahedra, it yields a macroscopic extensive degeneracy of the ground state. As shown by Ramirez and co-workers [41], both water ice and spin ice ground states have the same configuration entropy at $T = 0$, governed by the ice rules. Another interesting mapping of the spin ice behaviour can be made with the "two up-two down" configuration which minimizes the energy of a Heisenberg antiferromagnet on the pyrochlore lattice [42].

In spin ices and similar materials, the magnetic moments become correlated at temperatures corresponding to the energy scale of the exchange interactions (given by the Curie-Weiss constant θ_{CW}), but a transition towards LRO does not occur. The magnetic ground state shows a very peculiar local ordering, due to the existence of strong constraints on the free energy [43]. Local rules govern the spin configurations and states with zero magnetization must maximize the entropy. These constraints result in an emergent gauge

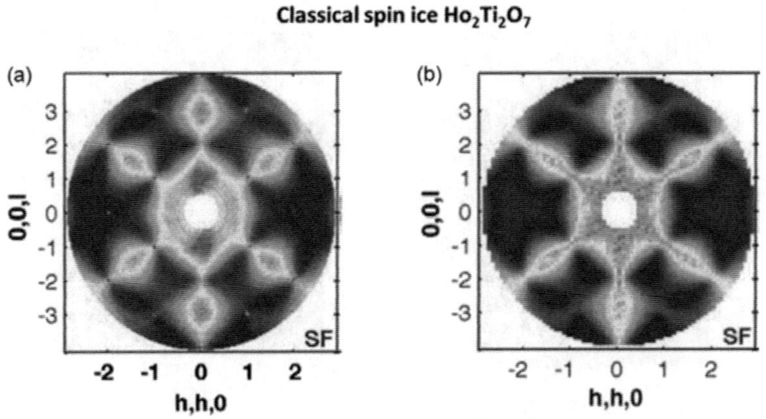

Figure 5. $Ho_2Ti_2O_7$ classical spin ice: a) experimental diffuse scattering map in a (110) plane measured by polarized neutrons in the spin flip channel; b): calculation in the dipolar spin ice model. From Ref. [44].

structure, (as in electrodynamics) and in a non-divergent magnetic field associated with the local rules governing the spin arrangements. As a consequence, the spatial variation of the spin correlations has an algebraic dependence similar to that of the dipolar interaction. In Fourier space, it corresponds to a very anisotropic shape of the diffuse scattering showing "pinch-points" or sharp bow-tie features [44].

The classical spin ice materials $Ho_2Ti_2O_7$ and $Dy_2Ti_2O_7$ which exist as large single crystals and possess large magnetic rare earth moments have enabled beautiful experiments probing this description. They are actually "dipolar spin ices", where the effective ferromagnetic first neighbour exchange is the sum of AF super-exchange and dipole-dipole interaction, the latter being screened for further neighbour distances. The strong Ising anisotropy coming from the crystal field scheme with ground state doublet well separated from the excited states constrains the spins along their local <111> axes. In this frame, the initial calculation of the neutron diffuse scattering maps by Monte Carlo simulation and comparison with the experimental maps was the first proof of the validity of dipolar spin ice model [40]. Later on, the use of polarized neutrons yielded a clear observation of the pinch points by isolating the "spin flip" cross section ([44] and Fig. 5), which entails most of magnetic modulations. In classical spin ice, flips of magnetic moments violating the ice rules can propagate along Dirac strings with low energy cost, leading to a separation of emergent magnetic charges akin to magnetic monopoles [45]. Such excitations are thermally activated and contribute to the diffuse scattering. Under magnetic field, the monopoles diffuse along the field and make random walk in the perpendicular plane, yielding specific features in the diffuse scattering, which were actually observed in $Dy_2Ti_2O_7$ [46].

From theory, starting from the classical spin ice and introducing transverse terms in the energy yields quantum spin ice behaviour, which allows tunnelling processes between spin configurations respecting the ice rules [47]. Accordingly, a unique quantum spin ice ground state with vanishing entropy at $T = 0$ could be built from a coherent superposition of large number of classical spin ice configurations. As a consequence of these quantum fluctuations, a new type of excitation has been predicted, quantum analogue of the monopoles, and associated with an emergent electromagnetic field. This "emergent photon" has not been observed up to now and constitutes a challenge of inelastic neutron scattering. Diffuse scattering experiments are also a crucial test for the quantum spin ice behaviour. The predicted diffuse scattering maps show the suppression of the pinch points by quantum fluctuations at

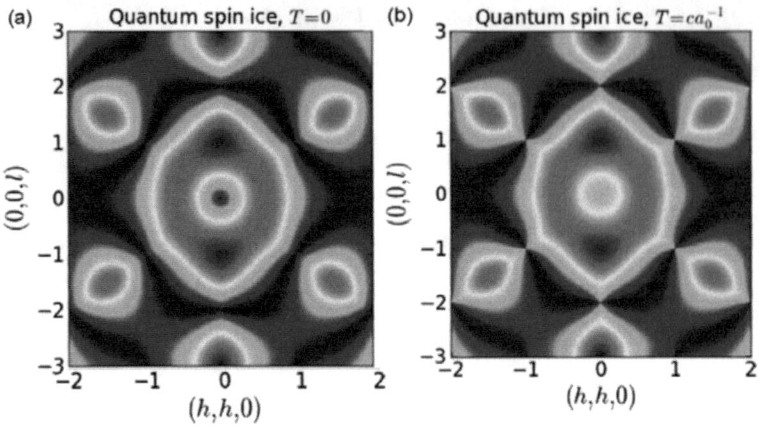

Figure 6. Predicted quantum spin ice behaviour as shown on diffuse scattering maps. a) at T = 0: the pinch points previously observed in classical spin ice are supressed by quantum fluctuations. In b) when increasing temperature, they are progressively restored by new excitations, called "emergent photons". From [47].

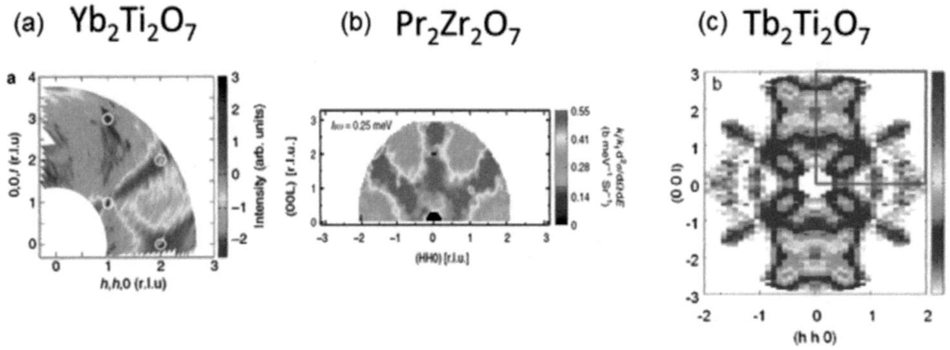

Figure 7. Single crystal neutron diffuse scattering maps in the (hh0, 00l) plane of the quantum spin ice candidates: a) $Yb_2Ti_2O_7$, from Ref. [48]; b) $Pr_2Zr_2O_7$, from Ref. [49]; c) $Tb_2Ti_2O_7$, from Ref. [50].

T = 0 and their progressive restoration with increasing temperature by the "emergent photons" (Fig. 6).

Experimentally, several routes towards quantum spin ice behaviours are possible, involving low magnetic moments, low energy crystal field levels, and/or quadrupolar interactions. Several pyrochlore materials such as $Yb_2Ti_2O_7$, $Tb_2Ti_2O_7$ and $Pr_2Zr_2O_7$ are presently considered as quantum spin ice candidates. None of them actually showed the behaviour predicted theoretically, but each of them offered an interesting and original variant to the classical spin ice behaviour, which can be seen on the diffuse maps at very low temperatures. In $Yb_2Ti_2O_7$, where Yb is a Kramers ion mapped to S=1/2, the polarized neutrons diffuse scattering maps show diffuse rods with features reminiscent of pinch points at <111> positions (Fig. 7a). This diffuse scattering becomes suppressed at $T_C = 0.2$ K when a first order transition settles in towards a ferromagnetic-like state, with low ordered magnetic moment and hysteretic features [48]. In $Pr_2Zr_2O_7$ where Pr is a non Kramers ion, similar broad features can be observed along the 111 directions, fully smearing the pinch points, and the modulations of the diffuse scattering occur in the inelastic channel, in a small energy range [49] (Fig. 7b).

$Tb_2Ti_2O_7$ spin liquid (Fig. 7c) offers the most complex and challenging alternative to classical spin ice behaviour. The diffuse scattering maps show well defined pinch points at the 111 positions, but with respect to $Ho_2Ti_2O_7$ and $Dy_2Ti_2O_7$ classical spin ices, the spectral weight is transferred out of the zone centre, and the intensity of the 200 diffuse peak explodes in a butterfly shape, with a dominant spectral weight on ½ ½ ½ and equivalent positions [50–52]. This antiferromagnetic version of the Coulomb phase suggests conservations rules holding over larger scales than a tetrahedron. Interestingly this disordered fluctuating ground state holds anisotropic excitations, which propagate perpendicular to the pinch points. Diffuse scattering combined with polarization and energy analysis of the transmitted neutron beam, yielded a clean determination of these exotic excitations, showing that their dispersion does not reflect that predicted for the emergent photon. The quantum spin ice state predicted by theoreticians therefore remains to be found.

6. Conclusion

This chapter summarizes some diffuse neutron scattering studies, focusing on local order in binary alloys and magnetic correlations in spin ices. These cases are only a few among the numerous fields of condensed matter physics which benefit from this approach. High luminosity spectrometers, polarization or energy analysis of the scattered neutrons, studies in extreme conditions of temperature, pressure, magnetic or electric field, should enable further studies in various domains, requiring sophisticated models and theories.

References

[1] J.M. Cowley, J. Appl. Phys. **21**, 24 (1950); B.E. Warren, X ray diffraction Courier Corp. (1969)
[2] B. Borie, C.J. Sparks, Acta Cryst. A **27**, 198 (1971)
[3] F. Frey, Mat. Sc. Engi. A. **294-296**, 178 (2000) QC; B. Toudic et al., Science **319**, 69 (2008)
[4] W. Marshall, S.W. Lovesey, Theory of thermal neutron scattering (Clarendon Press, 1971)
[5] S.M. Yusuf et al., Phys. Rev. B **82**, 094410 (2010)
[6] C. Ruegg et al., Phys. Rev. Lett. **93**, 037207 (2004)
[7] H. Yoshisawa et al., J. Phys. Soc. Jpn. Lett. SJ **57**, 3686 (1988)
[8] T. Fennell SFN **13**, 04001 (2014)
[9] `www.ill.eu/fr/instruments-support/instruments-groups/instruments/d7/description/`
[10] I.A. Blech, B.L. Averbach, Phys. Rev. B **137**, 1113 (1965); J. Davidowski et al., Neutron news **25**, 21 (2015)
[11] K.D. Rouse, M.J. Cooper Acta Cryst. A **26**, 682 (1970); A.W. Hewatt, Acta. Cryst. A **35**, 248 (1979)
[12] V.F. Sears, Neutron scattering lengths and cross sections tables, Neutron News **3**, 26 (1992)
[13] I. Mirebeau, M. Hennion, G. Parette Phys. Rev. Lett. **53**, 687 (1984); I. Mirebeau and G. Parette Phys. Rev. B **82**, 104203 (2010)
[14] M.E. Collins, Magnetic Critical Scattering; Oxford Series on Neutron Scattering in Condensed Matter
[15] E. Burkel et al., Z. Phys. B **35**, 227 (1979)
[16] S.E. Dutton et al., Phys. Rev. Lett. **108**, 187206 (2012)

[17] F. Bley, P. Cenedese, S. Lefebvre, AIP Conference Proceedings **89**, 276 (1982)
[18] E. Kentzinger et al., Phys. Rev. B **61**, 14975 (2000)
[19] V. Pierron-Bohnes et al., Phys. Rev. B **51**, 5760 (1995)
[20] T. Mehaddene et al., Eur. Phys. J. B **41**, 207 (2004)
[21] T. Priem et al., J. de Phys. **50**, 2217 (1989)
[22] R.A. Medina, J.W. Cable Phys. Rev. B **15**, 1539 (1977); H.R. Child and J.W. Cable Phys. Rev. B **13**, 227 (1976)
[23] J.E. Greedan et al., Phys. Rev. B **79**, 014427 (2009)
[24] A. Apetrei et al., Phys. Rev. Lett. **97**, 206401 (2006)
[25] I.V. Solovyev Phys. Rev. B **67**, 174406 (2003)
[26] J.E. Greedan et al., Phys. Rev. B **43**, 5682 (1991); J.N. Reimers Phys. Rev. B **46**, 193 (1992)
[27] FullProf, `http://www.ill.eu/sites/fullprof`
[28] P. Bordet SFN **9**, 139 (2008)
[29] B.A. Frandsen, X. Yang, S.J. Billinge, Acta Crystallogr. A **70**, 3 (2014)
[30] J.A.M. Paddison, A.L. Goodwin, Phys. Rev. Lett. **108**, 017204 (2012)
[31] R. Kikuchi, Phys. Rev. **81**, 988 (1951); J.M. Sanchez, V. Pierron-Bohnes and F. Mejia-Lira Phys. Rev. B **51**, 3429 (1995)
[32] S.T. Bramwell, M.J.P. Gingras Science **294**, 5546 (2001)
[33] M.J.P. Gingras, P.A. Mac Clarty, Rep. Prog. Phys. **77**, 056501 (2014)
[34] F. Ducastelle, F. Gautier, J. Phys. F **6**, 2039 (1976)
[35] M.Yu. Lavrentiev et al., Phys. Rev. B **75**, 014208 (2007)
[36] P. Erhart, A. Caro, M. Serrano de Caro, B. Sadigh, Phys. Rev. B **77**, 134206 (2008)
[37] P.C. Clapp, S.C. Moss, Phys. Rev. **171**, 754 (1968)
[38] M. Hennion, J. Phys. F **13**, 2351 (1983)
[39] A.V. Ruban, P.A. Korzhavyi, B. Johansson, Phys. Rev. B **77**, 094436 (2008)
[40] S. Bramwell et al., Phys. Rev. Lett. **87**, 047207 (2011)
[41] A.P. Ramirez et al., Nature (London) **399**, 333 (1999)
[42] R. Moessner, J.T. Chalker, Phys. Rev. B **58**, 12049 (1998)
[43] S.V. Isakov et al., Phys. Rev. Lett. **93**, 167204 (2004); C.L. Henley, Phys. Rev. B **71**, 014424 (2005)
[44] T. Fennell et al., Nat. Phys. 3, 566 (2007); Science **326** 415 (2009)
[45] C. Castelnovo, R. Moessner, S.L. Sondhi, Nature **451**, 42 (2008); I.A. Ryzhkin, M.A. Ryzhkin, JETP Lett. **93**, 384 (2011); C.L. Henley, Phys. Rev. B **71**, 014424 (2005)
[46] D.J.P. Morris et al., Science **326**, 411 (2009)
[47] O. Benton, O. Sikora, N. Shannon, Phys. Rev. B **86**, 075154 (2012)
[48] L.J. Chang et al., Nat. Comm. **3**, 992 (2012)
[49] K. Kimura et al., Nat. Comm. **4**, 1934 (2013)
[50] S. Petit et al., Phys. Rev. B **86**, 174403 (2012); S. Guitteny et al., Phys. Rev. Lett. **111**, 087201 (2013)
[51] T. Fennell et al., Phys. Rev. Lett. **109**, 017201 (2012)
[52] K. Fritsch et al., Phys. Rev. B **89**, 094410 (2013)

EPJ Web of Conferences **155**, 00007 (2017) DOI: 10.1051/epjconf/201715500007
JDN 22

Inelastic neutron scattering

Sylvain Petit[a]

Laboratoire Léon Brillouin, CEA-CNRS, Université Paris-Saclay, CE-Saclay, 91191
Gif-sur-Yvette, France

Abstract. The goal of the JDN22 school was to propose a progressive teaching eager to improve the expertise of students in *neutron diffraction*. Neutron-based techniques have indeed proved for decades to be essential tools in the investigation of condensed matter. This lecture is however concerned with *inelastic neutron scattering* and is thus somehow apart. In the context of this school, it should then only be considered as a brief introduction. We give simple examples along with the basics of the spectrometers, and finally useful formula for the inelastic cross sections in different situations. We strongly encourage interested readers to refer to the bibliography for more detailed information.

1. Introduction

Condensed matter physics has benefited tremendously from both elastic and inelastic neutron scattering techniques for more than half a century. Since then, these techniques have been used to study a wide variety of materials. The main reasons arise from the unique physical properties of the neutron itself, zero electric charge, spin half... Also, since the energies of the neutrons with "thermal" wavelengths match the energy scales of many condensed matter systems, it is possible to use them to probe the dynamics. Excitations that can be studied via neutron scattering range in energy from a few meV to a fraction of an electron volt. As a result, inelastic neutron scattering provides crucial information about the energetic terms that appear in a model Hamiltonian. While interesting from a pure fundamental point of view, it should be emphasized that these terms also determine the response of these systems to an external perturbation, hence allowing one to predict the functional capabilities of materials.

To give the flavor and introduce the basics of inelastic neutron scattering, it is useful to describe first simple examples, either in the context of lattice dynamics or of spin dynamics.

1.1 Lattice dynamics

Let's consider a single mono-atomic chain sketched in Fig. 1(a). Here, the atoms with mass M are at distances a, and bound together by springs (constant Γ). In other words, the Hamiltonian describing this system is of the form:

$$\mathcal{H} = \sum_n \frac{1}{2} M \mathbf{v}_n^2 + \frac{1}{2}\Gamma(\mathbf{x}_{n+1} - \mathbf{x}_n - a)^2$$

$\mathbf{v}_n$ and $\mathbf{x}_n$ are the velocity and position of the individual mass at site labeled n. A diffraction experiment on such a sample (if it would exist!) would give Bragg spots at positions such that

[a] e-mail: `sylvain.petit@cea.fr`

$Q_n = \frac{2\pi}{a}\, n$, characterizing the spatial periodic arrangement of the atoms. Information about the energy scales of the problem, such as inertia M and springs Γ is clearly absent.

To get more insight into this question, it is useful to examine the excited states of the chain. Writing the atomic position as $\mathbf{x}_n = \mathbf{u}_n + n\ \mathbf{a}$, where the $n\ a$ are the equilibrium positions, classical mechanics shows that the u_n displacements of the atoms are governed by the equations of motion:

$$M\frac{d^2\mathbf{u}_n}{dt^2} = \Gamma(\mathbf{u}_{n+1} - \mathbf{u}_n - \mathbf{u}_n + \mathbf{u}_{n-1})$$

whose solutions are plane waves with energy $\omega_{\mathbf{k}}$ and wave-vector $\mathbf{k}$. Introducing these solutions into the above equation, we get the dispersion relation:

$$-M\omega_{\mathbf{u}}^2 = \Gamma\left(e^{\,i\mathbf{ka}} + e^{-i\mathbf{ka}} - 2\right)$$

hence

$$\omega_{\mathbf{k}} = 2\sqrt{\frac{\Gamma}{M}}\left|\sin\frac{\mathbf{ka}}{2}\right| \tag{1}$$

This dispersion is a periodic function of $\mathbf{k}$, goes to zero at the Brillouin zone centers and reaches extrema at the zone boundaries. From this simple calculation, we conclude that the excited states of the chain consist in harmonic displacements of the atoms. Energies (frequencies) and wavevectors (wavelength) are related by $\omega_{\mathbf{k}} = 2\sqrt{\frac{\Gamma}{M}}\left|\sin\frac{\mathbf{ka}}{2}\right|$. Quantification of these modes leads to the concept of "phonons", a set of harmonic oscillators, characterized by the energies $\omega_{\mathbf{k}}$ and polarization $\mathbf{e}_{\mathbf{k}}$ (that correspond to the direction of vibrations of the atoms). The displacements $\mathbf{u}_n$ are linear combinations of the different phonon modes:

$$\mathbf{u}_n(t) = \sum_k e^{\,i(n\mathbf{ka} - \omega_{\mathbf{k}} kt)} \frac{\hbar}{\sqrt{2M\omega_{\mathbf{k}}}}(a_{\mathbf{k}} + a_{-\mathbf{k}}^{+})\,\mathbf{e}_{\mathbf{k}}$$

where the $\{a_{\mathbf{k}}, a_{\mathbf{k}}^{+}\}$ are the annihilation and creation phonon operators.

At this point, the question that naturally arises is what would be the outcome of an inelastic neutron scattering experiment on that chain? The response is simple: it would allow one to observe directly the dispersion relation $\omega_{\mathbf{k}}$. Practically, as we will see later, inelastic neutron scattering instruments allow one to define trajectories in the $(\mathbf{k}, \omega)$ space; the number of neutrons recorded in the detector is essentially zero along such trajectories except when they cross the $\omega_{\mathbf{k}}$ map. As shown in Fig. 1(a–d), a series of constant energy or constant Q cuts will, step by step, reconstruct the dispersion. Figure 1f shows an example of such a mapping obtained in the case of PbTe [1] by gathering cuts together: the dispersion becomes visible, above some background. Note that the intensity along the dispersion is not constant and depends on $\mathbf{k}$.

Back to the example of the chain, fitting Eq. (1) to the measured $\omega_{\mathbf{k}}$ would allow to measure Γ. In some sense, this reasoning is very similar to what we do when we want to analyze diffraction data. Here also the positions or relative orientations of the spins are "measured" based on a model that best fits the diffraction pattern.

1.2 Spin dynamics

Let's now consider the example of the two sublattices antiferromagnet depicted in Fig. 2(a). From diffraction alone, the relative orientation of spins is easily determined, but the exchange that couples two neighboring spins, along with the amplitude of a single ion anisotropy term, would be impossible to guess. From inelastic neutron scattering, however, taking advantage of

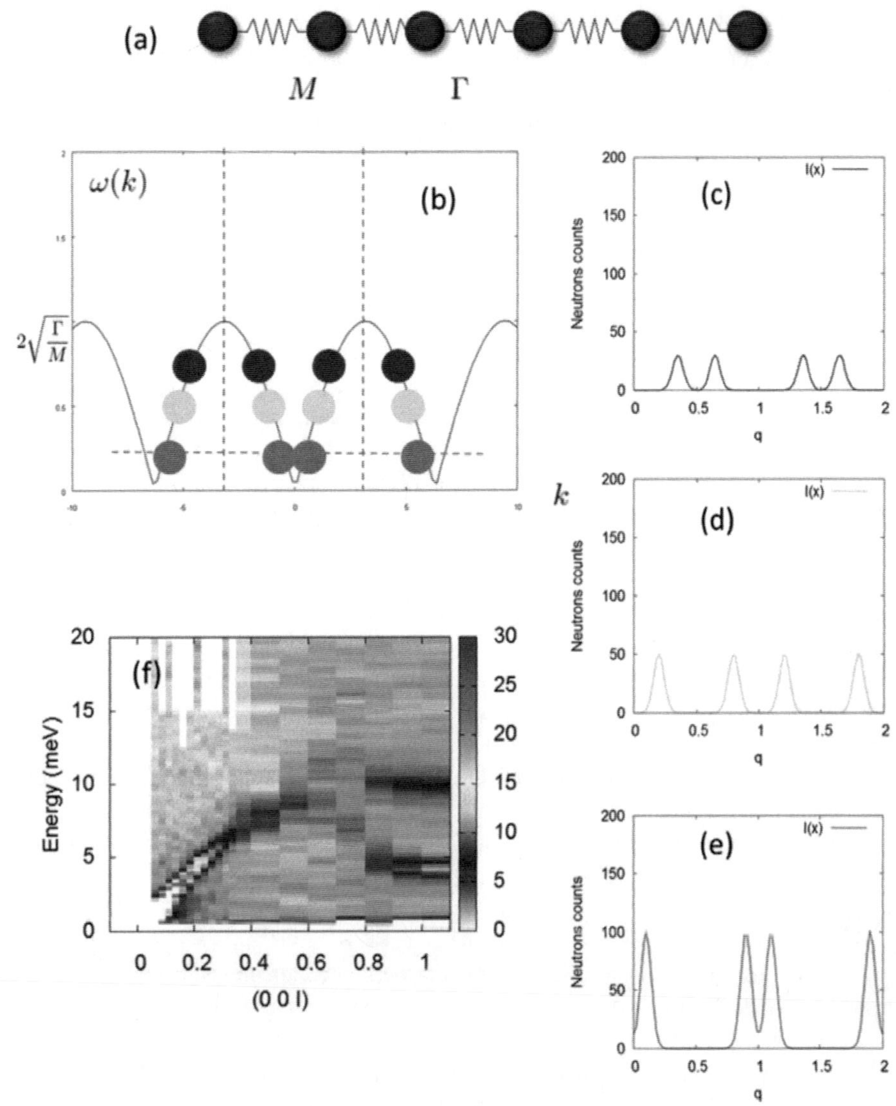

Figure 1. Lattice dynamics of a one 1D chain. (a) Chain model. (b) Shows the dispersion relation, while (c–e) show constant-energy scans across this dispersion. Each peak correspond to a different crossing point in (b). (f) Presents a map of longitudinal phonon propagating along $\mathbf{c}^*$ in PbTe, obtained by gathering constant-Q scans together.

the magnetic interaction, it is possible to determine the dispersion of elementary excitations of such a magnet. The latter, called spin waves, are collective excitations that physically correspond to precession motions of the spins around the average direction. From theory, based on the Heisenberg model:

$$\mathcal{H} = \sum_{i,j} J\mathbf{S}_i.\mathbf{S}_j + D(S_i^z)^2$$

we learn that this dispersion is given by:

$$\omega_{\mathbf{k}} = S\sqrt{(6J + D)^2 - J^2(2\cos k_x a + 2\cos k_y a + 2\cos k_z a)^2}$$

Figure 2(b) and (c) illustrate these dispersions respectively without and with a single ion anisotropy D that forces the spins to lie down along . In close analogy with the phonon case, a series of constant energy or constant Q cuts will, step by step, reconstruct this dispersion. Fitting it to the above equation will provide numerical values of the exchange J and of the anisotropy D. Note especially the presence of a spin gap for $D \neq 0$. Figure 2(d) shows an example of such a mapping obtained in the case of the multiferroic $YMnO_3$ by gathering such cuts together: as for PbTe, the dispersion becomes visible, above some background. Note also that the intensity along the dispersion is not constant and depends on **k**.

This theory assumes that there is a long range ordering (a phase transition towards a Néel state). In analogy with field theory, the "spin waves" are (bosonic) $\Delta S = 1$ particles and the Néel state is considered as the "vacuum" from which these particles emerge. It should be stressed, however, that a number of magnets never order, owing to strong quantum fluctuations, low dimensionality and frustration effects. For instance, recent progress in the physics of low dimensional systems have shown that spin waves are actually made of two spin-half quantum entities called spinons, which under usual circumstances remain confined and thus not visible. Actually, this concept of confinement, when particles are bound together by an interaction whose strength increases with increasing particle separation, is a fundamental idea of modern physics. The most widely studied model in which this phenomenon takes place is the spin-half Heisenberg antiferromagnetic chain with first-neighbor interactions. In agreement with the Wagner and Mermin theorem, quantum fluctuations prevent static long-range order and the ground state remains a spin liquid, with power-law decaying antiferromagnetic correlations. In such a chain, spinons can be seen as domain walls between reversed antiferromagnetic domains; they can move freely and independently along the chain, as no energy cost is involved: in other words, they do not confine. In this case, the magnetic response shows no well-defined spin wave excitations but an incoherent spectrum called two-spinons continuum, which reflects the range of energies corresponding to the different spinon pairs (see Fig. 2(e)).

2. Principles

We now turn to the description of spectrometers dedicated to inelastic neutron scattering experiments [2–10]. As for diffraction, most of those experiments demand a monochromatic beam, to be prepared from the white beam produced by the neutron source (reactor or pulsed source). The main idea is then to determine the final energy of the neutrons by analyzing the scattered beam. Taking the energy difference between incident and outgoing neutrons gives the energy $E_i - E_f = \omega$ given to the sample (note that this quantity can be negative, which correspond to a situation where the neutrons receive energy from the sample and are thus accelerated).

There are essentially two methods for discerning the energy of the neutrons (see Figs. 3 and 4): the first one is by measuring the time it takes for neutrons to travel the distance between the sample to the detector, known as time-of-flight neutron spectroscopy (TOF) [9]. This method assumes that the incident beam is chopped to define the time reference, hence reducing the effective average flux on the sample.

The second method is by Bragg scattering from a single crystal, known as triple-axis spectroscopy (TAS) [8, 10, 11]:

$$\lambda = \frac{2\pi}{|\mathbf{k}_f|} = 2d \sin\theta$$

It takes advantage of the continuous flux of the source. It was invented by Brockhouse in the mid 1950s at Chalk River [7]. Although there have been many improvements, the principle behind this technique has not changed from that originally developed. The term triple-axis

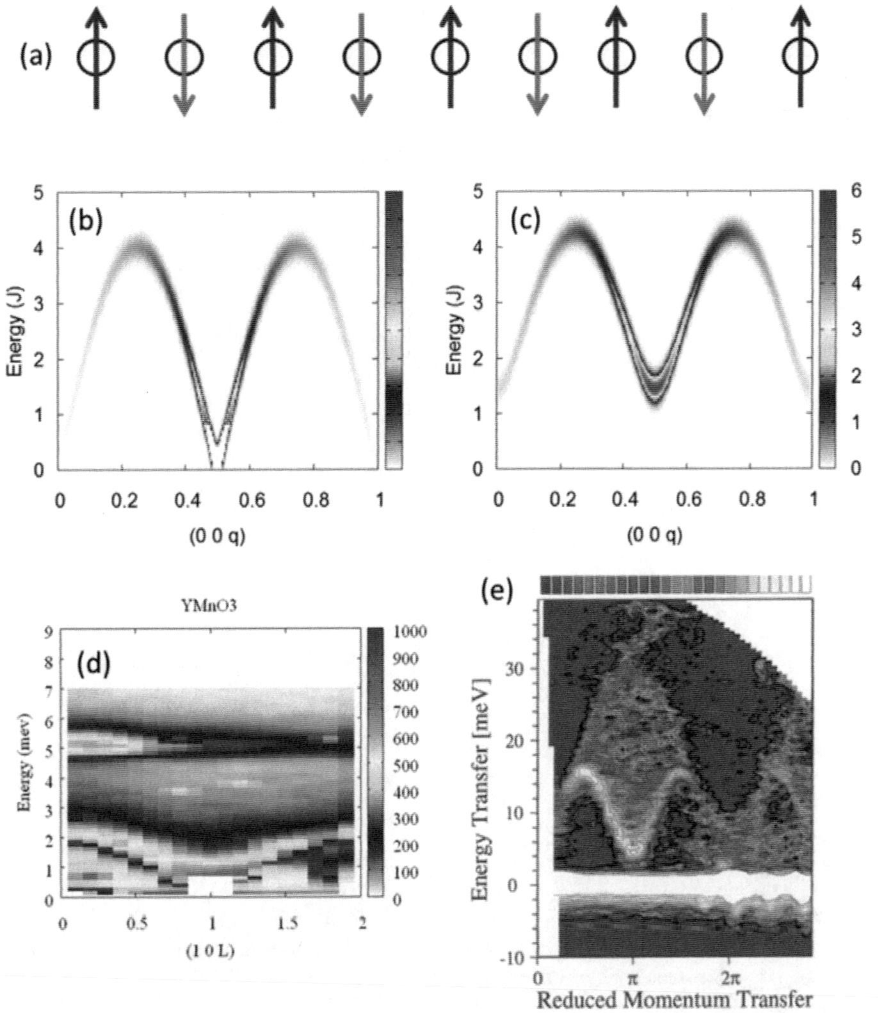

Figure 2. Spin dynamics in a simple two sublattices antiferromagnet. (a) displays the antiferromagnetic chain. (b-c) Show the neutron scattering expected intensity in two different cases, without and with single ion anisotropy respectively. (d) Presents a map of the dispersion of spin waves in the multiferroic material $YMnO_3$ and (e) a map of the two spinons continuum in a 1D chain (M. Arai, M. Fujita, M. Motokawa, J. Akimitsu, and S.M. Bennington, Phys. Rev. Lett. **77**, 3649 1996).

derives from the fact that neutrons are scattered from three crystals as they travel from the source to the detector.

Depending on the possible processes that take place in the sample, the scattered neutrons may have a different energy from the incident one. This is sketched in Fig. 3(a) using rays of several colors. At this point, the conservation of energy and momentum writes:

$$E_i - E_f = \omega$$

$$\mathbf{k}_i - \mathbf{k}_f = \mathbf{Q}$$

with neutron incident and final energies and momentum related by:

$$E_{i,f} = \frac{\hbar^2 k_{i,f}^2}{2M_n}$$

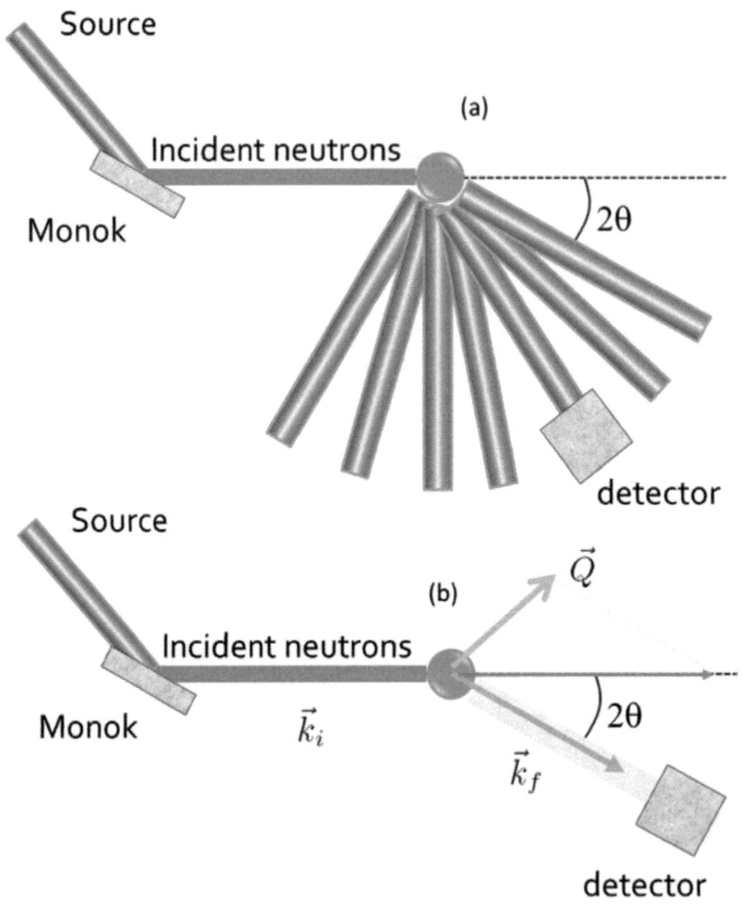

Figure 3. Sketches of a neutron scattering experiment. (a) Pointing out that under a given scattering angle, the beam contains neutrons of different energies; (b) illustrates the conservation of energy and momentum and (c) the rotation necessary to let coincide **Q** defined by $\mathbf{k}_i - \mathbf{k}_f$ and the reciprocal space of the sample.

Importantly, we notice that the scattering angle $\alpha = (\mathbf{k}_i, \mathbf{k}_f)$ is determined by the modulus of the wavevector **Q**:

$$\cos\alpha = \frac{k_i^2 + k_f^2 - Q^2}{2k_i k_f}$$

The wavevector **Q** is determined, for the moment, by $\mathbf{k}_i - \mathbf{k}_f$, and has no special meaning in terms of reduced reciprocal coordinates (h, k, ℓ) of the sample itself. This becomes only true if two particular directions of the sample, within the scattering plane, are known. In that case, $\mathbf{Q} = \mathbf{k}_i - \mathbf{k}_f$ can be projected onto the sample's reciprocal lattice, so that (h, k, ℓ) are uniquely defined. Rotating the sample around the vertical axis allows to address a new set of (h', k', ℓ') coordinates, corresponding to the same modulus.

Each of these methods, TAS or TOF (see Fig. 4(a) and (b) respectively) is most useful in studying a particular type of problem. Traditionally, TAS has been the preferred method for studying single crystals, and TOF for studying powders. Yet, in principle, nothing forbids to measure a single crystal on a TOF instrument. It is just necessary to rotate the sample around the vertical axis, to "reconstruct", for each scattering angle and each time of flight value, the (h, k, ℓ, ω) coordinates. Today, the main advantage of TAS spectrometer consists in the fact

DOI: 10.1051/epjconf/201715500007

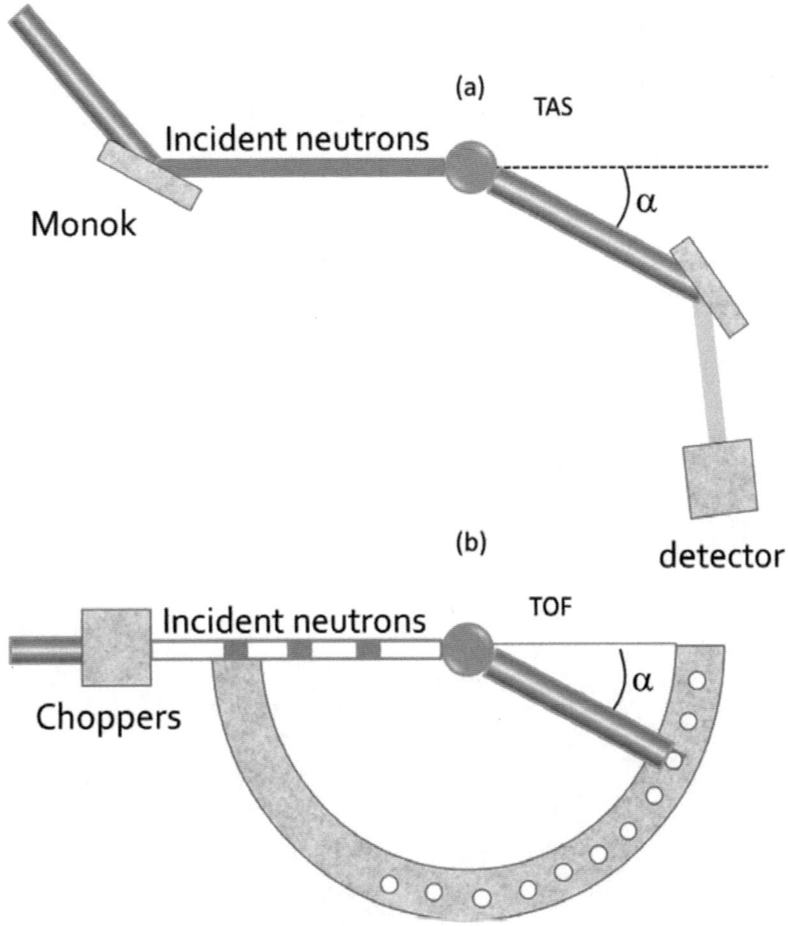

Figure 4. Sketch of TAS and TOF spectrometers.

that it is quite easy to vary external parameters like temperature, magnetic field, pressure or electric field and to measure the response in a small region of $(\mathbf{Q}, \omega)$ space. In contrast, TOF techniques are useful to obtain a survey of excitations in a rather extended $(\mathbf{Q}, \omega)$ region.

2.1 TAS in practice

As is clear from the Bragg law, monochromators and analysers select k_o, $2k_o$, $3k_o$ ... (k_o standing for k_i or a k_f). Even if the measured intensities are normalized to the same number of incident neutrons with the help of a low sensibility detector placed in the incident beam, called "monitor", it remains important to correct these data for those "bad" neutrons with $2k_o$, $3k_o$ ... This supposes that we know the proportion of harmonics in the incident beam, a quantity that especially depends on the incident energy, the kind of source (thermal, hot, cold, see Fig. 5(a)), and also on the installation of the spectrometer on a neutron guide or not. This correction, called "monitor correction", is shown in Fig. 5(b).

Another issue arising because of harmonics is the possible contamination of the measurement by *elastic* processes ($\omega = 0$) occurring in the sample provided that:

$$nk_o = mk_f \ \ (n, m) \text{ integers}$$

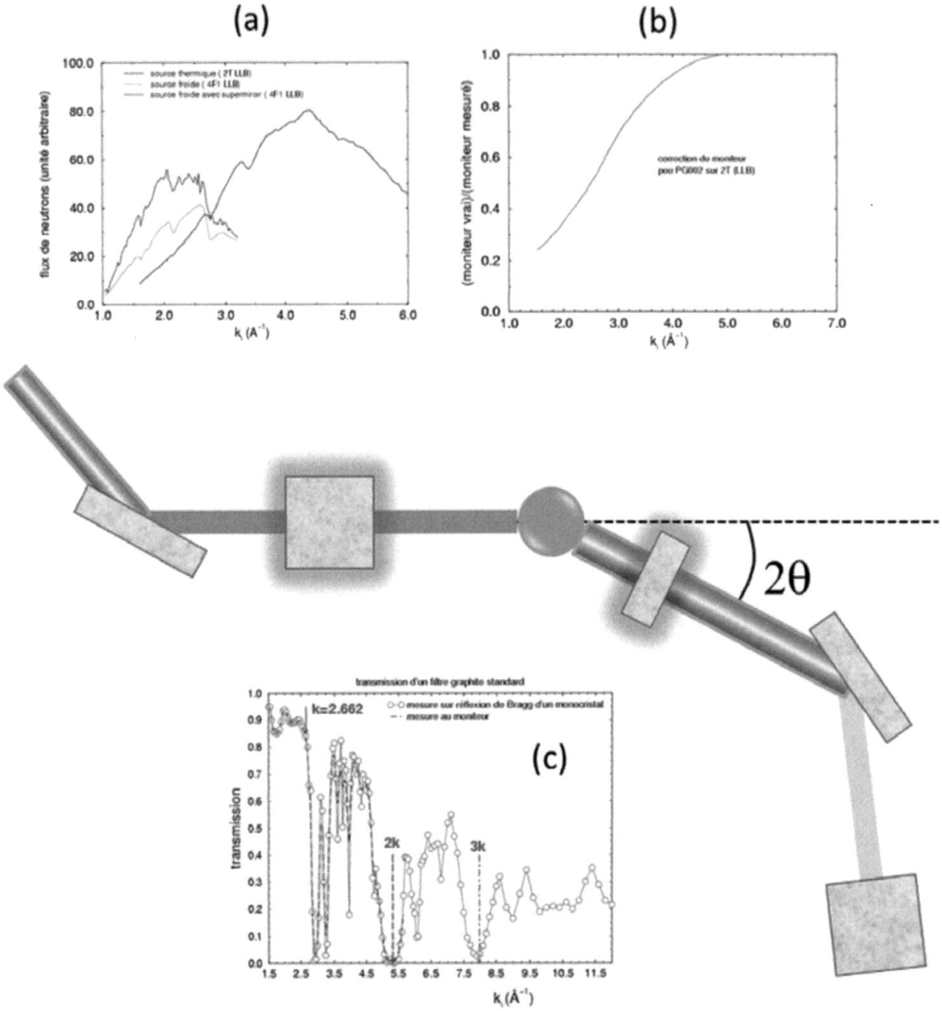

Figure 5. Monitor and filters in a TAS experiment. (a–b) show the incident flux, depending on the kind of source, and the monitor correction obtained for 2T [10]. (c) illustrates the filtering with pyrolithic graphite, high transmission for k_o and almost zero for the harmonics.

The intensity of such processes can be extremely large compared to the "true" inelastic expected signal. To partly eliminate this problem, filters are used, either placed in the incident or in the scattered beam. The idea is to use a peculiarity of the transmission of pyrolithic Graphite, which is almost 1 for a given k_f and 0 for the harmonics, at least (in practice) for $2k_f$. This happens for "magic" values, that are widely used on TAS, 4.1, 3.85, 2.662, 2.57, 1.97, 1.64, and 1.48 Å^{-1} (see Fig. 5(c)). The Beryllium is another material widely used on cold TAS. If cooled at 70 K, its transmission is about 1 for wavevectors lower than 1.55 Å^{-1} and 0 above.

Finally, it is of fundamental importance to introduce the concept of resolution. Actually, a spectrometer cannot isolate a single point in (Q, ω) space; rather, it integrates over a small volume around that point, an effect which can be modeled by the following convolution:

$$I(\mathbf{Q}, \omega) = \int d\mathbf{Q}'\, d\omega'\; R(\omega - \omega', \mathbf{Q} - \mathbf{Q}')\; S(\omega', \mathbf{Q}')$$

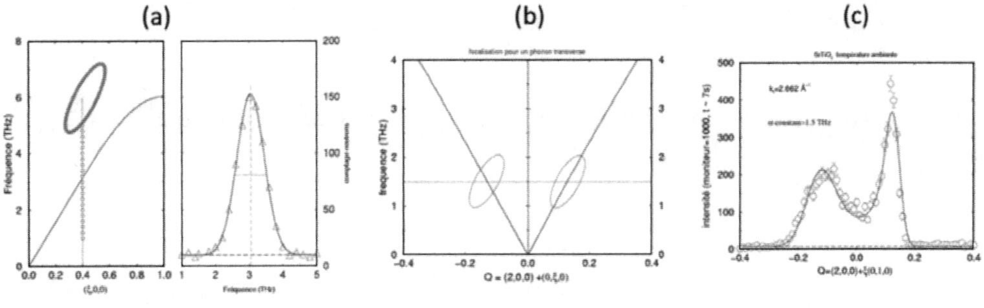

Figure 6. Illustration of the ellipsoid resolution on a TAS [10]. The effect of the inclination of the long axis with respect to the slope of the dispersion is especially illustrated. If these are parallel, then the signal is narrow and intense, while it is broad and weak in the opposite situation. These two possibilities are called focused and unfocused conditions.

with

$$R(\omega, \mathbf{Q}) = V_i\, V_f\, \exp\left[-\{\mathbf{Q}, \omega\}^T \bar{A}\, \{\mathbf{Q}, \omega\}\right]$$

$\{\mathbf{Q}, \omega\}$ is a 4 dimensional vector with components $\mathbf{Q}$ and ω. $\bar{A}$ is a 4×4 matrix. Explicated in its eigen-directions, becomes a nice 4D-ellipsoid with long and short axes, called resolution ellipsoid. A constant energy scans across the dispersion relation gives en enlarged peak as sketched in Fig. 6(a). There exists softwares able to calculate R [10] but we shall give here only guidelines:

- the z-axis can be considered as decoupled from the other quantities. The vertical resolution is generally quite poor.
- because of the inclination of the ellipsoid in (q_x, q_y, ω) space, the analysis of dispersion can become tricky, with especially a focused and an unfocused side: as sketched in Fig. 6(b) and (c), the apparent width of branches can look very different.
- the smaller k_f or k_i, the better the resolution; however, since $V_{i,f} \approx k_{i,f}^3$, the less intense is the expected signal.

3. Cross sections

The quantity that we ultimately measure is the neutron cross section. It is proportional to the number of neutrons scattered in a solid angle Ω with a final energy E_f.

3.1 Nuclear interaction

In the case of nuclear interaction, this cross section can be written as follows:

$$\frac{\partial^2\sigma}{\partial\Omega\partial E_f} = \frac{k_f}{k_i} \sum_{i,j} b_i\, b_j \int_{-\infty}^{+\infty} dt\, \langle e^{\,i\, \mathbf{Q}\, \mathbf{R}_i(t=0)}\, e^{-i\, \mathbf{Q}\, \mathbf{R}_j(t)} \rangle\, e^{-i\omega t}$$

where b_i is the scattering length of ion at site i with time dependent position $\mathbf{R}_i(t)$.

3.1.1 The frozen lattice

Moving in gradual steps, let's first assume that the lattice is frozen, so that these positions do not depend on time. The vector $\mathbf{R}_i$ is decomposed into a vector that labels the unit cell $\mathbf{R}_m^o$ and a second vector within the unit cell $\mathbf{r}_\ell$. Concomitantly, the site i is decomposed into two new indexes $i = (m, \ell)$.

$$\mathbf{R}_i(t) = \mathbf{R}^o_m + \mathbf{r}_\ell$$

Note that because of the spatial periodicity, the b_i's only depend on ℓ. The cross section becomes:

$$\begin{aligned}
\frac{\partial^2\sigma}{\partial\Omega\partial E_f} &= \frac{k_f}{k_i}\,\delta(\omega)\sum_{m,n,\ell,\ell'} b_\ell\, b_{\ell'}\, e^{\,i\,\mathbf{Q}\,(\mathbf{R}^o_m+\mathbf{r}_\ell-\mathbf{R}^o_n-\mathbf{r}_{\ell'})} \\
&= \delta(\omega)\sum_{m,n} e^{\,i\,\mathbf{Q}\,(\mathbf{R}^o_m-\mathbf{R}^o_n)} \sum_{\ell,\ell'} b_\ell\, b_{\ell'}\, e^{\,i\,\mathbf{Q}\,(\mathbf{r}_\ell-\mathbf{r}_{\ell'})} \\
&= \delta(\omega)\sum_{m,n} e^{\,i\,\mathbf{Q}\,(\mathbf{R}^o_m-\mathbf{R}^o_n)}\,|F_N(\mathbf{Q})|^2 \\
&= \delta(\omega)\sum_{\tau} \delta(\mathbf{Q}-\tau)\,|F_N(\mathbf{Q})|^2
\end{aligned}$$

where τ is a reciprocal lattice vector and $F_N(\mathbf{Q})$ is the familiar nuclear structure factor:

$$F_N(\mathbf{Q}) = \sum_\ell b_\ell\, e^{\,i\,\mathbf{Q}\,\mathbf{r}_\ell}$$

As a result, in this approximation, the cross section is purely elastic ($\omega = 0$) and consists of Dirac functions at the nodes of the reciprocal lattice. Their intensities are given by the squared modulus of the structure factor.

3.1.2 The harmonic approximation

We now turn to the "vibrating" lattice approximation. The $\mathbf{R}_i(t)$ become:

$$\mathbf{R}_i(t) = \mathbf{R}^o_m + \mathbf{r}_\ell + \mathbf{u}_{m,\ell}(t)$$

where $\mathbf{u}_{m,\ell}(t)$ describes the harmonic oscillations, in terms of phonon modes, of the atom labeled (m, ℓ). The cross section writes:

$$\frac{\partial^2\sigma}{\partial\Omega\partial E_f} = \frac{k_f}{k_i}\sum_{m,n} e^{\,i\,\mathbf{Q}\,(\mathbf{R}^o_m-\mathbf{R}^o_n)}\sum_{\ell,\ell'} b_\ell\, b_{\ell'} e^{\,i\,\mathbf{Q}\,(\mathbf{r}_\ell-\mathbf{r}_{\ell'})}\int_{-\infty}^{+\infty} dt \langle e^{\,i\,\mathbf{Q}\,\mathbf{u}_{m,\ell}}\, e^{-i\,\mathbf{Q}\,\mathbf{u}_{n,\ell'}(t)}\rangle\, e^{-i\omega t}$$

Using quantum properties of the phonon creation and annihilation operators, the expectation value $\langle..\rangle$ can be rearranged:

$$\langle e^{\,i\,\mathbf{Q}\,\mathbf{u}_{m,\ell}}\, e^{-i\,\mathbf{Q}\,\mathbf{u}_{n,\ell'}(t)}\rangle = e^{-W_\ell-W_{\ell'}}\, e^{\langle \mathbf{Q}\,\mathbf{u}_{m,\ell}\;\mathbf{Q}\,\mathbf{u}_{n,\ell'}(t)\rangle}$$

W_ℓ is the so called Debye-Waller factor, that physically describes the average squared deviations of the ions at temperature T with respect to their equilibrium position:

$$W_\ell = \frac{1}{2}\left\langle [\mathbf{Q}\,\mathbf{u}_{m,\ell}]^2\right\rangle$$

The cross section becomes:

$$\frac{\partial^2\sigma}{\partial\Omega\partial E_f} = \frac{k_f}{k_i}\sum_{m,n} e^{\,i\,\mathbf{Q}\,(\mathbf{R}^o_m-\mathbf{R}^o_n)}\sum_{\ell,\ell'} b_\ell\, b_{\ell'}\, e^{\,i\,\mathbf{Q}\,(\mathbf{r}_\ell-\mathbf{r}_{\ell'})}\, e^{-W_\ell-W_{\ell'}}\int_{-\infty}^{+\infty} dt\; e^{\langle \mathbf{Q}\,\mathbf{u}_{m,\ell}\;\mathbf{Q}\,\mathbf{u}_{n,\ell'}(t)\rangle}\, e^{-i\omega t}$$

Expanding the exponential function to second order, one obtains:

$$e^{\langle \mathbf{Q}\,\mathbf{u}_{m,\ell}\;\mathbf{Q}\,\mathbf{u}_{n,\ell'}(t)\rangle} \approx 1 + \langle \mathbf{Q}\,\mathbf{u}_{m,\ell}\;\mathbf{Q}\,\mathbf{u}_{n,\ell'}(t)\rangle\; + \;...$$

The zeroth order leads to a slight modification of the structure factor which now incorporates the Debye-Waller factor:

$$\frac{\partial^2\sigma}{\partial\Omega\partial E_f} = \delta(\omega)\sum_\tau \delta(\mathbf{Q}-\tau)\,|F(\mathbf{Q})|^2$$

$$F_N(\mathbf{Q}) = \sum_\ell b_\ell\, e^{\,i\,\mathbf{Q}\,\mathbf{r}_\ell} e^{-W_\ell}$$

The second order term is the phonon cross section:

$$\frac{\partial^2\sigma}{\partial\Omega\partial E_f} = \sum_s A_s\left[\left(1+n(\omega_{\mathbf{Q},s})\right)\,\delta(\omega-\omega_{\mathbf{Q},s}) + n(\omega_{\mathbf{Q},s})\,\delta(\omega+\omega_{\mathbf{Q},s})\right]$$

and consists in a sum of Dirac functions on the different phonon branches. This finally demonstrates that an inelastic neutron scattering experiment allows a direct measurement of the dispersion relations. The intensity along the dispersion is given by the "dynamical" structure factor:

$$A_s = |F_s|^2$$

with

$$F_s(\mathbf{Q}) = \sum_\ell b_\ell\, e^{\,i\,\mathbf{Q}\,\mathbf{r}_\ell}\, e^{-W_\ell}\sqrt{\frac{\hbar}{M_\ell\omega_{q,s}}}\,(\mathbf{Q}.\mathbf{e}_{q,\ell,s})$$

The scalar product between the wavevector $\mathbf{Q}$ and the polarization $\mathbf{e}_{q,\ell,s}$ is especially useful as it allows to distinguish transverse and longitudinal vibrations. Longitudinal phonons correspond to vibrations parallel to the propagation vector $\mathbf{q}$, while transverse phonons correspond to vibrations perpendicular to $\mathbf{q}$. To this end, it suffices to measure in different Brillouin zones. With $\mathbf{Q} = \tau + \mathbf{q}$ (τ is the zone center), different configurations can be chosen so that $\mathbf{Q}$ is approximatively parallel or perpendicular to a given polarization. Figure 7(a) shows an example of this trick in the case of CaF_2, a sample very often used for practicals at LLB: maps obtained for transverse (Fig. 7(c)) and longitudinal (Fig. 7(d)) phonons propagating along $(hh0)$ are shown as well as a typical individual Q-scan used to produce the maps.

Note that positive energy processes correspond to the creation of a phonon by a neutron, while negative energy process correspond to the annihilation of a phonon. $n(\omega_{\mathbf{Q},s})$ is the Bose number at temperature T:

$$n(\omega_{\mathbf{Q},s}) = \frac{1}{e^{\frac{\hbar\omega_{\mathbf{Q},s}}{k_B T}} - 1}$$

Further, the two processes do not have the same intensity: the creation is always possible, even at low temperature, while the annihilation supposes that the initial quantum state of the sample contains phonons. For instance, the annihilation is forbidden at very low temperature since there are no phonons in the ground state configuration. This is called the "detailed

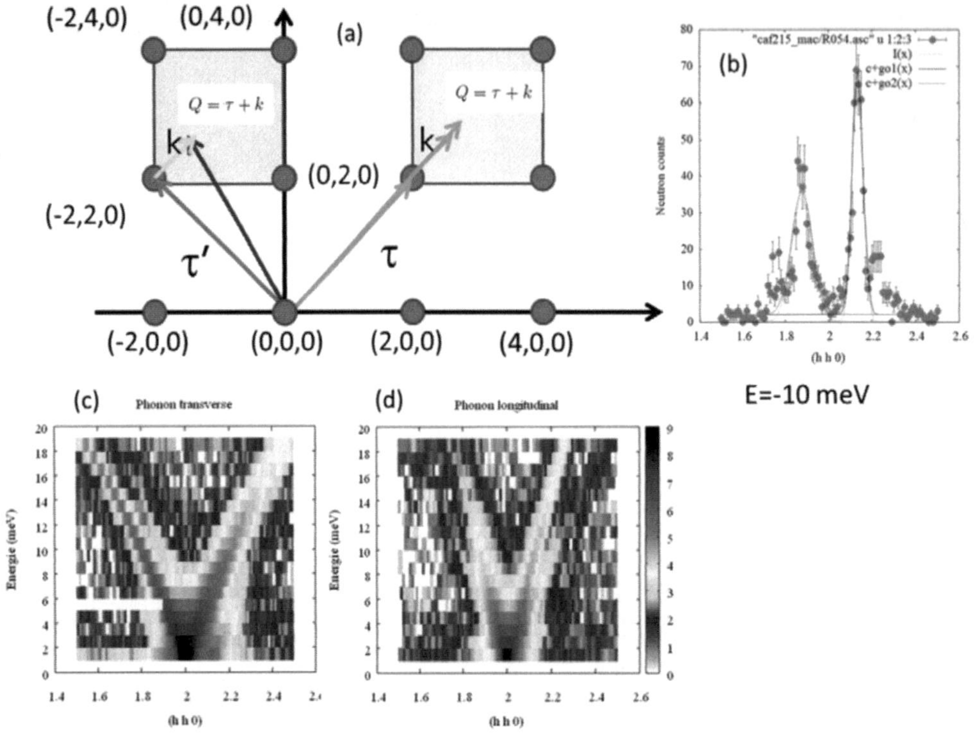

Figure 7. Inelastic neutron scattering tutorial in CaF_2. (a) Sketch of the reciprocal plane ($hk0$), along with two Brillouin zone appropriate fir the study of transverse and longitudinal phonon modes. (b) Shows an individual constant-energy scan while (c) and (d) show the maps obtained by gathering such scans in transverse and longitudinal conditions.

balance"; this is an extremely important concept, also known in other spectroscopy methods like IR or Raman scattering for instance.

3.1.3 Connection with diffraction

As already pointed out, a diffraction experiment do not analyze the final energy of the neutrons but rather integrates over these final energies. As a result, this kind of measurement is sensitive to both the pure elastic Bragg scattering, but also to the excitations. As the cross section is very small, it appears as a kind of diffuse scattering, which can be difficult to distinguish from the background. Note that ω cannot be larger than the incident energy E_i, so that the corresponding cross section reads:

$$\frac{\partial \sigma}{\partial \Omega} = \int_{-\infty}^{E_i} d\omega \, \frac{k_f}{k_i} \sum_{i,j} b_i \, b_j \int_{-\infty}^{+\infty} dt \, \langle e^{\, i \, \mathbf{Q} \, \mathbf{R}_i} \, e^{-i \, \mathbf{Q} \, \mathbf{R}_j(t)} \rangle \, e^{-i\omega t}$$

If E_i is large enough compared to the dynamics, one usually integrates out the ω dependence to write:

$$\frac{\partial \sigma}{\partial \Omega} = \sum_{i,j} b_i \, b_j \, \langle e^{\, i \, \mathbf{Q} \, \mathbf{R}_i(t=0)} \, e^{-i \, \mathbf{Q} \, \mathbf{R}_j(t=0)} \rangle$$

In other words, diffraction measures instantaneous correlations. Diffraction at zero energy measures elastic (or infinite time) correlations; inelastic scattering measures the whole (finite time) spectrum of these correlations.

3.2 Magnetic interaction

In the case of the magnetic interaction, the cross section is given by:

$$\frac{\partial^2\sigma}{\partial\Omega\partial E_f} = \frac{k_f}{k_i}(\gamma_N r_o)^2 \sum_{i,j}\int_{-\infty}^{+\infty} dt\, \langle \mathbf{S}_{\perp,i}\, \mathbf{S}_{\perp,j}(t)\, e^{\,i\,\mathbf{Q}\,\mathbf{R}_i}\, e^{-i\,\mathbf{Q}\,\mathbf{R}_j(t)}\rangle\, e^{-i\omega t}$$

$\gamma_N = 1.913$ is the gyromagnetic ratio of the neutron, r_o is the classical electron radius. $\mathbf{S}_\perp$ is the component of the spin perpendicular to the scattering vector $\mathbf{Q}$. It is worth noting that:

$$\mathbf{S}_{\perp,i}\, \mathbf{S}_{\perp,j}(t) = \sum_{\alpha,\beta} S_{\alpha,i}\left(\delta_{\alpha,\beta} - \frac{Q_\alpha\, Q_\beta}{Q^2}\right) S_{\beta,j}(t)$$

The exact derivation of the cross section goes well beyond this lecture; we shall however give the final result:

$$\frac{\partial^2\sigma}{\partial\Omega\partial E_f} = \frac{k_f}{k_i}(\gamma_N r_o)^2 \sum_{m,n} e^{i\,\mathbf{Q}\,(\mathbf{R}_m^o - \mathbf{R}_n^o)} \sum_{\ell,\ell'} f_\ell(\mathbf{Q}) f_{\ell'}^*(\mathbf{Q}) e^{i\mathbf{Q}(\mathbf{r}_\ell - \mathbf{r}_{\ell'})} e^{-W_\ell - W_{\ell'}}$$

$$\int_{-\infty}^{+\infty} dt\, \langle \mathbf{S}_{\perp,\ell}\, \mathbf{S}_{\perp,\ell'}(t)\rangle\, e^{-i\omega t}$$

3.3 Frozen spins

If the spins are frozen, $\mathbf{S}_{\perp,\ell}$ do not depend on time, and we are left with a pure static ($\omega = 0$) contribution:

$$\frac{\partial^2\sigma}{\partial\Omega\partial E_f} = (\gamma_N r_o)^2\, \delta(\omega) \sum_{\tau_m} \delta(\mathbf{Q} - \tau_m)\, |\mathbf{F}_m(\mathbf{Q})|^2$$

τ_m is a reciprocal magnetic lattice vector and $\mathbf{F}_m$ is the magnetic structure factor:

$$\mathbf{F}_m(\mathbf{Q}) = \sum_\ell \mathbf{S}_{\perp,\ell}\, e^{\,i\,\mathbf{Q}\,\mathbf{r}_\ell}\, e^{-W_\ell}\, f_\ell(\mathbf{Q})$$

3.4 Spin waves

In the spin wave approximation, the elementary excitations, called spin waves, correspond to precession motions of the spins around the magnetization direction [12–17]. These spin waves are free particles carrying a spin 1. Tough calculations show that:

$$\frac{\partial^2\sigma}{\partial\Omega\partial E_f} = \sum_s A_s[\left(1 + n(\omega_{\mathbf{Q},s})\right)\delta(\omega - \omega_{\mathbf{Q},s}) + n(\omega_{\mathbf{Q},s})\delta(\omega + \omega_{\mathbf{Q},s})]$$

In close analogy with the case of phonons, the cross section takes the form of a sum of Dirac functions on the different spin waves branches. Again, this formula shows that an inelastic neutron scattering experiment allows a direct measurement of the dispersion

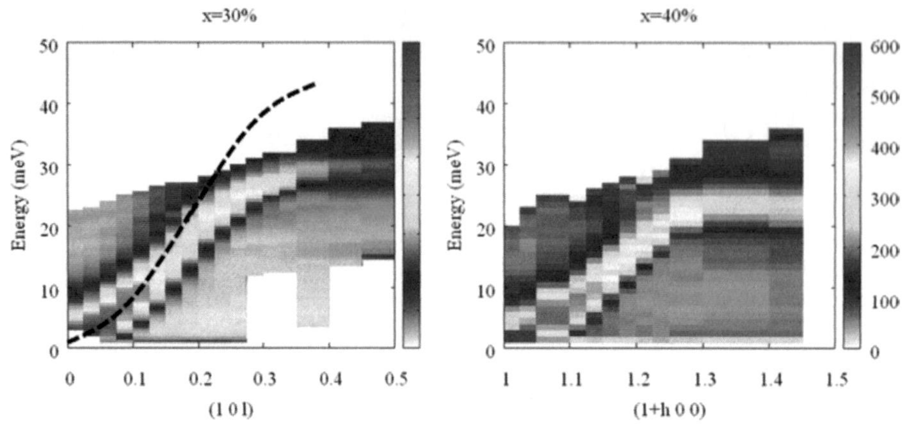

Figure 8. Spin wave dispersion in the cubic ferromagnet $La_{1-x}Sr_xMnO_3$ compounds. A simple cosine dispersion is expected. Note the flattening close to the zone boundary which has not been explained yet despite a huge theoretical work.

relations. The intensity of the modes is given by a "dynamical" structure factor A_s, whose general expression, however, is rather complex; its derivation goes far beyond the scope of this lecture. Figure 8 shows the mapping measured for instance in the $La_{1-x}Sr_xMnO_3$ compounds, very well know for their magneto-resistance properties. These systems are simple cubic ferromagnets (lattice parameter a), and one expects a dispersion relation reading as:

$$\omega_{\mathbf{k}} = 2JS(3 - \cos k_x a - \cos k_y b - \cos k_z a)$$

As can be observed from Fig. 8, this simple form fits very well the experiment at small **q**, but is clearly not adapted close to the zone boundary: the dispersion becomes rather flat, an effect which remains mysterious so far, not reproduced by theoretical calculations. This illustrates the relevance and simplicity of inelastic neutron scattering, which definitely provides valuable information about the spectrum of elementary excitations even in very complex condensed matter.

3.5 Crystal field levels

The response due to the inelastic scattering by crystal field excitations is also very useful in practice. The latter correspond to the different transition between states describing the electronic configurations of the (usually 4f or 5f) electrons. Denoting E_n and ϕ_n the energies and wave functions, we get:

$$\int_{-\infty}^{+\infty} dt\ \langle \mathbf{S}_{\perp,\ell}\ \mathbf{S}_{\perp,\ell}(t)\ \rangle\ e^{-i\omega t} = \int_{-\infty}^{+\infty} dt\ \sum_{n,m} \frac{e^{-E_n/T}}{Z} \langle\phi_n|\mathbf{S}_{\perp,\ell}|\phi_m\rangle\ \langle\phi_m|\mathbf{S}_{\perp,\ell}|\phi_n\rangle\ e^{\,i(E_m - E_n - \omega)t}$$

$$= \sum_{n,m} \frac{e^{-E_n/T}}{Z} |\langle\phi_n|\mathbf{S}_{\perp,\ell}|\phi_m\rangle|^2 \delta(E_m - E_n - \omega)$$

$$Z = \sum_n e^{-E_n/T}$$

from which one deduces:

$$\frac{\partial^2\sigma}{\partial\Omega\partial E_f} = \frac{k_f}{k_i}(\gamma_N r_o)^2 \sum_{\ell} |f_\ell(Q)|^2 \, e^{-2W_\ell} \sum_{n,m} \frac{e^{-E_n/T}}{Z} |\langle\phi_n|\mathbf{S}_{\perp,\ell}|\phi_m\rangle|^2 \delta(E_m - E_n - \omega)$$

As a result, the scattering due to crystal field shows Dirac functions at all possible transitions between two eigenstates $\omega = E_m - E_n$. Note that there is no particular Q dependence: indeed, no correlations are expected since the moments do not interact with each other, except the Debye Waller and the form factor. At low temperature, most of the possible transitions disappear since $\frac{e^{-E_n/T}}{Z} \approx \delta_{n,1}$, so that we are left with the transition from the ground state only. The cross section becomes:

$$\frac{\partial^2\sigma}{\partial\Omega\partial E_f} = \frac{k_f}{k_i}(\gamma_N r_o)^2 \sum_{\ell} |f_\ell(Q)|^2 \, e^{-2W_\ell} \sum_{m} |\langle\phi_1|\mathbf{S}_{\perp,\ell}|\phi_m\rangle|^2 \delta(E_m - E_1 - \omega)$$

3.6 Metal

In a metal, or by the way, in any material where the elementary excitations are fermions carrying a spin half (like spinons in 1D chains, for instance), the spin excitations is composed of a continuum that is nothing but a kind of two particles density of states. Indeed, since the particles carry a spin half, two "flips", hence a pair of particles, is necessary to match the $\Delta S = 1$ transition of the neutron spin. For a given $(\mathbf{k}, \omega)$, different pairs of fermions contribute provided their energies $E_{\mathbf{q}}$ and $E_{\mathbf{q}'}$ fulfill the condition $\omega = E_{\mathbf{q}'} + E_{\mathbf{q}}$ and provided the momentum is conserved $\mathbf{k} = \mathbf{q} + \mathbf{q}'$. The corresponding spectral weight depends however on details of the Hamiltonian [18].

4. Fluctuation dissipation

We would like now to close this lecture by pointing out the link between two equivalent languages: correlations and response function or generalized susceptibility [19]. Actually, the response of a system under a time dependent perturbation $B(t)$ results in the emergence of an observable A. Applying a magnetic field will for instance result in a magnetization, or applying a force will result in atomic displacements. In the framework of linear response theory, A and B are "proportional", related by a generalized susceptibility G:

$$A(t) = \int dt' G(t - t') B(t')$$

or in frequency representation:

$$A(\omega) = G(\omega) B(\omega)$$

where the Fourier transform $G(\omega)$ is defined as:

$$G(\omega) = \lim_{\epsilon \to 0^+} \int dt e^{i\omega t - \epsilon t} G(t)$$

The theory shows that [19]:

$$G(t) = i\theta(t)\langle[A, B(t)]\rangle_o$$

where the expectation value is to determined with the unperturbed Hamiltonian of the system $(B = 0)$. If one denotes E_n and ϕ_n the eigen-energies and wave functions of the system, one obtains:

$$G(t) = i\theta(t)\sum_n \frac{e^{-E_n/T}}{Z}\langle\phi_n|\,[A, B(t)]\,|\phi_n\rangle_o$$

Introducing $\langle\phi_n|A|\phi_m\rangle = A_{nm}$, and calculating the commutator, this becomes:

$$G(t){=}i\,\theta(t)\sum_{n,m}\frac{e^{-E_n/T}}{Z}(A_{nm}\,B_{mn}\,e^{\,i(E_m-E_n)t}{-}B_{nm}\,A_{mn}\,e^{-i(E_m-E_n)t})$$

or

$$\begin{aligned}
G(t) &= i\theta(t)\sum_{n,m}\frac{e^{-E_n/T}}{Z}\,A_{nm}\,B_{mn}\,e^{\,i(E_m-E_n)t} - \frac{e^{-E_m/T}}{Z}\,B_{mn}A_{nm}\,e^{-i(E_n-E_m)t} \\
&= i\theta(t)\sum_{n,m}\frac{e^{-E_n/T}-e^{-E_m/T}}{Z}\,A_{nm}\,B_{mn}\,e^{\,i(E_m-E_n)t} \\
&= i\theta(t)\sum_{n,m}\frac{e^{-E_n/T}}{Z}\left(1-e^{-(E_m-E_n)/T}\right)\,A_{nm}\,B_{mn}\,e^{\,i(E_m-E_n)t}
\end{aligned}$$

Using:

$$\begin{aligned}
\lim_{\epsilon\to0^+}\int dte^{\,i\omega t-\epsilon t}\,i\theta(t)\,e^{\,ixt} &= \lim_{\epsilon\to0^+}\frac{i[0-1]}{i(\omega+x)-\epsilon} \\
&= \lim_{\epsilon\to0^+}\frac{-1}{(\omega+x)+i\epsilon} \\
&= -\mathrm{Pf}\frac{1}{\omega+x}+i\pi\delta(\omega+x)
\end{aligned}$$

We have for the imaginary part of $G(\omega)$:

$$\begin{aligned}
G''(\omega) &= \pi\sum_{n,m}\frac{e^{-E_n/T}}{Z}\left(1-e^{-(E_m-E_n)/T}\right)A_{nm}\,B_{mn}\,\delta(\omega+(E_m-E_n)) \\
&= \pi\left(1-e^{-\omega/T}\right)\sum_{n,m}\frac{e^{-E_n/T}}{Z}\,A_{nm}\,B_{mn}\,\delta(\omega+(E_m-E_n))
\end{aligned}$$

On the other hand, correlations between A and B are defined by:

$$C(t) = \langle AB(t)\rangle_o$$

which can be rewritten as:

$$C(\omega) = \sum_{n,m}\frac{e^{-E_n/T}}{Z}A_{nm}\,B_{mn}\delta(\omega+(E_m-E_n))$$

As a result:

$$G(\omega) = \pi\left(1-e^{-\omega/T}\right)\,C(\omega)$$

$$C(\omega) = \frac{1}{\pi}\frac{1}{1-e^{-\omega/T}}\,G(\omega)$$

These calculations demonstrate the relations between the generalized susceptibility (also called response function) and the correlations. $\left(1 - e^{-\omega/T}\right)^{-1}$ is called the detailed balance factor. The magnetic inelastic neutron cross section is a measure of correlations between spins; it is also nothing but the imaginary part of the frequency dependent magnetic susceptibility.

5. Conclusion

In this lecture, the basics of inelastic neutron scattering were presented. We have especially pointed out that diffraction measures instantaneous correlations, while inelastic scattering determine the time dependence of those correlations. This allows a direct measurement of the dispersion of elementary excitations, like phonons or spin waves.

References

[1] W. Szuszkiewicz, unpublished
[2] G.L. Squires, *Introduction to the Theory of Neutron Scattering* (Dover Publications, 1996)
[3] P.G. de Gennes, *Theory of Neutron Scattering by Magnetic Crystals*, in *Magnetism*, Vol. III, Eds. G.T. Rado and H. Suhl (New York: Academic Press, 1963), p.115–147
[4] Neutron Scattering, ed. by K. Sköld, D.L. Price, *Methods in Experimental Physics*, Vol. 23C (Academic Press, New York, 1987)
[5] S.W. Lovesey, *Theory of Neutron Scattering from Condensed Matter*, Vol 2 (Oxford Science Publications, 1984)
[6] Y.A. Izyumov and N.A. Chernoplekov, *Neutron Spectroscopy* (Plenum, 1994)
[7] B.N. Brockhouse. *Inelastic scattering of neutrons in solids and liquids*, Vienna. 1961. p. 113; http://nobelprize.org/nobel_prizes/physics/laureates/1994/brockhouse-lecture.html
[8] G. Shirane, S.M. Shapiro and J. Tranquada, *Neutron Scattering with a Triple-Axis Spectrometer*, Basic Techniques (Cambridge University Press, 2002)
[9] *Diffusion inélastique de neutrons par temps de vol.* J. Ollivier et J.-M. Zanotti. JDN 16, Collection SFN **10**, 379–423 (2010) https://doi.org/10.1051/sfn/2010006
[10] *La diffusion inélastique des neutrons sur monocristal. Le spectromètre 3-axes.* B. Hennion. JDN 16, Collection SFN **10**, 357–378 (2010) https://doi.org/10.1051/sfn/2010005
[11] Z. Yamani, Z. Tun, and D.H. Ryan, Can. J. Phys. **88**, 771–797 (2010)
[12] Magnetic Excitations, W.G. Stirling et K.A. McEwen, *Methods of Experimental Physics*, Vol. 23 Part C (Academic Press, 1987)
[13] D.C. Mattis, *Theory of Magnetism I* (Springer Verlag, 1988)
[14] R.M. White, *Quantum Theory of Magnetism* (Springer Verlag, 1987)
[15] A. Auerbach, *Interacting electrons and Quantum Magnetism* (Springer Verlag, 1994)
[16] D.C. Wallace, PR **128**, 1614 (1962) Spin waves in complex lattice
[17] *Numerical simulations and magnetism.* S. Petit, JDN 18, Collection SFN **12**, 105–121 (2011) https://doi.org/10.1051/sfn/201112006
[18] *Magnétisme de bande et Supraconductivité: l'apport de la diffusion inélastique des neutrons.* Yvan Sidis, JDN 20, Collection SFN **13**, 04003 (2014) https://doi.org/10.1051/sfn/20141304003
[19] *Physique statistique hors d'équilibre: Processus irréversibles linéaires.* N. Pottier (EDP Sciences, 2007)

AUTHOR INDEX